W9-AXO-442

In Praise of TOMATOES

Tasty Recipes, Garden Secrets, Legends & Lore

Ronni Lundy

RECIPES BY CHEF JOHN STEHLING OF EARLY GIRL EATERY

GARDEN SECTION BY BARBARA CILETTI

LARK BOOKS

A Division of Sterling Publishing Co., Inc.
New York

Art Director: **DANA M. IRWIN**

Photographer: **SANDRA STAMBAUGH**

Cover Designer: **BARBARA ZARETSKY**

Illustrator: **DANA M. IRWIN**

Assistant Art Director: **LANCE WILLE**

Assistant Editor: **NATHALIE MORNU**
REBECCA LIM

Editorial Assistance: **DELORES GOSNELL**

To Bill Best & his tomatoes

Library of Congress Cataloging-in-Publication Data

Lundy, Ronni.
In praise of tomatoes : tasty recipes, garden secrets, legends & lore / by Ronni Lundy ; recipes by John Stehling ; garden section by Barbara Ciletti.—1st ed.
p. cm.
Includes index.
ISBN 1-57990-421-1 (hardcover)
1. Cookery (Tomatoes) 2. Tomatoes. I. Stehling, John. II. Ciletti, Barbara J. III. Title.
TX803.T6L86 2004
641.6'5642—dc22 2003023510

10 9 8 7 6 5 4 3 2 1

First Edition

Published by Lark Books, a division of
Sterling Publishing Co., Inc.
387 Park Avenue South, New York, N.Y. 10016

Distributed in Canada by Sterling Publishing,
c/o Canadian Manda Group, One Atlantic Ave., Suite 105
Toronto, Ontario, Canada M6K 3E7

Distributed in the U.K. by Guild of Master Craftsman Publications Ltd.,
Castle Place, 166 High Street, Lewes, East Sussex, England
BN7 1XU Tel: (+ 44) 1273 477374, Fax: (+ 44) 1273 478606,
Email: pubs@thegmcgroup.com, Web: www.gmcpublications.com

Distributed in Australia by Capricorn Link (Australia) Pty Ltd.,
P.O. Box 704, Windsor, NSW 2756 Australia

A Note About Suppliers

Usually, the supplies you need for making the projects in Lark books can be found at your local craft supply store, discount mart, home improvement center, or retail shop relevant to the topic of the book. Occasionally, however, you may need to buy materials or tools from specialty suppliers. In order to provide you with the most up-to-date information, we have created a listing of suppliers on our Web site, which we update on a regular basis. Visit us at www.larkbooks.com, click on "Craft Supply Sources," and then click on the relevant topic. You will find numerous companies listed with their web address and/or mailing address and phone number.

If you have questions or comments about this book, please contact:
Lark Books
67 Broadway Asheville, NC 28801
(828) 253-0467

Manufactured in China

ISBN: 1-57990-421-1

TABLE OF CONTENTS

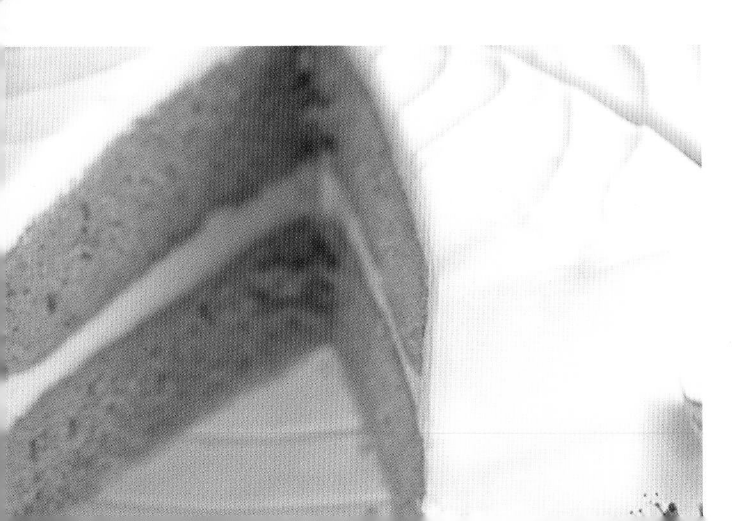

INDEX OF

Appetizers and Drinks

Salads and Dressings

Soups and Stews

Sandwiches

Stuffed Tomatoes

RECIPES

Main Courses

Side Dishes

Desserts

Breads

Condiments, Preserves, and Sauces

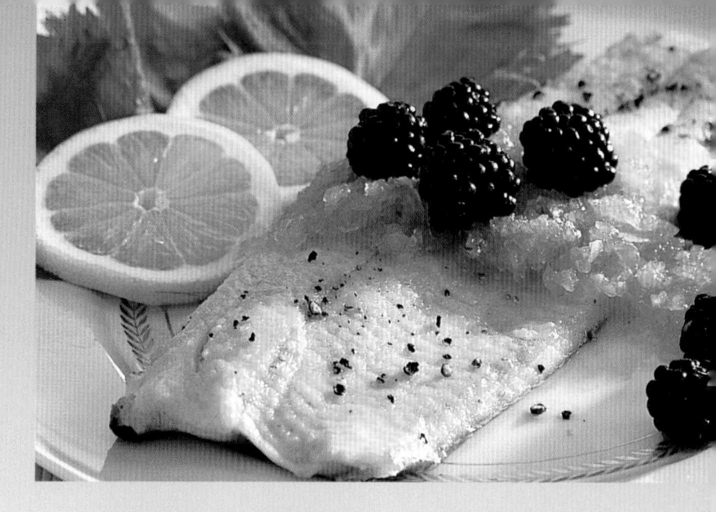

Look for me to find the best tomato recipe ideas for your table!

INTRODUCTION

Warm skin, smooth and thin as the finest silk, yields to the teeth with a gentle crack. Spurt of juice so abundant it cannot be contained by lips and so runs down the chin. Taste both tangy and sweet. Hidden chambers filled with seed and gel for the tongue to explore. Hues from scarlet red to milky white, coral sunset, sunburst orange, lemon yellow, green, purple, candy stripes. Is that a tiny mound of salt cupped in your palm? Dip in the tomato's edge ever so lightly, take the next bite, close your eyes and sigh.

Few sensual pleasures on earth are both so deeply satisfying and so readily accessible as that of eating a warm-from-the-vine tomato. Perhaps that explains why it is far and away the most popular of all home garden plants, and why it's *de rigeur* on salad plates and sandwiches.

And it's not just fresh tomatoes that we crave. Cooked tomatoes are the foundation of the two best-selling jarred condiments in the world, salsa and ketchup. Simmered slowly into sauce, tomatoes top pizza, enchiladas, meatloaf, and even bagels. Steamed and pureed, blended with milk, they become cream of tomato soup, the very definition of comfort.

The tomato is not only about the pleasures of eating, either. A chameleon in the kitchen (where it can be utilized as both vegetable and fruit), the tomato has also led a double life without. Rotten, it was once a highly favored object to be flung at politicians and/or actors who displeased us; ripe, it has been a much beloved subject immortalized in art, both fine and kitsch. The word has been used since the early 1900s to denote young women of decided pulchritude, but tomato or tomato can are also terms used to describe a lousy boxer, the sort sure to hit the canvas with a splat before the third round.

Townships across the globe celebrate the tomato with yearly festivals and vie to be known as the producer of the world's finest specimens.

It once was taken all the way to the U.S. Supreme Court to determine its botanical designation.

And what is so enticing about the tomato that thousands of young men and women from around the world each year make a pilgrimage to the narrow streets of tiny Bunol, Spain, to be gleefully pelted by a truckload of overripe ones?

Since its earliest days of cultivation and consumption, the tomato has been the focus of both desire and celebrity. Long ago it was rumored by some to be a great aphrodisiac, by others to be deadly, and apocryphal stories abound concerning the first person to eat one publicly. Myths have risen up about it, laws have been passed regarding it, cuisines have been built around it, and acres have been cleared to grow it.

As far as the garden is concerned, the last few years have seen nothing less than a tomato renaissance. Sure, corporate farm production and worldwide shipping virtually guarantee that you

IN *PRAISE OF TOMATOES* INTENDS TO BRING THAT ULTIMATE TOMATO EXPERIENCE TO YOU ON THE PAGE. WHAT YOU HOLD IS A BOOK THAT CELEBRATES THE MANY ASPECTS OF THIS FRUIT IN ALL OF ITS TANGY GLORY.

can find a fresh tomato (or general facsimile thereof) anywhere, at any time, from the supermarket bin to the corner convenience store. Yet even so, the demand for tomato seeds and plants for the home garden and small farm has increased not only in volume but also in diversity. In little more than a decade, the passion for a richer, truer, tastier tomato has seen an amazing boom in the production of once neglected heirloom tomatoes. Gardeners and consumers alike have gone to great lengths to save nearly lost tomato varieties–in a panoply of shapes, sizes, colors, and dispositions–all in search of the ultimate tomato experience.

In Praise of Tomatoes intends to bring that ultimate tomato experience to you on the page. What you hold is a book that celebrates the many aspects of this fruit in all of its tangy glory. At the heart of the matter (or perhaps that should be belly?) are some four dozen recipes featuring fresh, cooked, canned, and even fried tomatoes, in a variety of roles, from super star to essential supporting act.

The creator of those recipes, chef John Stehling, so loves the tomato that when he and his wife, Julie, opened a new southern-style restaurant in Asheville, North Carolina, they named it after one. The Early Girl Eatery concentrates on dishes made with the best local produce. In and around the Blue Ridge Mountains, farms and gardens have long produced tomatoes of every taste, size, and color, and John makes fine use of this bounty. John's plate-proven and tummy-tested recipes show up throughout the book, tantalizing tidbits for your tomato journey. You can use the handy recipe index on page 8 as well, to help you find your favorites when it's time to cook.

Also joining in this celebration of all things tomato is Barbara Ciletti, long-time garden writer from Colorado, who admits to an absolute passion for tomatoes. She has created The Essential Tomato Garden Primer on page 60. Complete with a chart of some 50 tomato varieties and their characteristics, the primer is full of information for creating and nurturing the tomato garden at home, whether you have a half-acre patch or a spot of sunlight on the terrace.

But as you know, the tomato is not simply for growing and eating. It also has a rich history and a culture that is fascinating, quirky and distinctly its own. You'll find all of that and more as you join us here, *in praise of tomatoes.*

What better way to raise a toast to the tomato than with that classic cocktail, the Bloody Mary?

The Consummate Bloody Mary

Serves 6

4 cups/1 L tomato juice

3 teaspoons horseradish

1 teaspoon Worcestershire sauce

1 teaspoon hot pepper sauce

juice of 1/2 lemon or lime

salt and pepper to taste

ice

vodka

6 cleaned, trimmed celery stalks with leaves

- Pour juice and all ingredients except vodka and ice into a covered jar or cocktail shaker and shake vigorously to combine.
- When ready to serve, fill a tall glass with ice for each person. Pour 1 jigger of vodka (or amount you prefer) into each glass. Insert a celery stick into the side of each and fill with Bloody Mary mix.

This classic eye-opening cocktail is a great way to serve tomatoes any time!

THE NAME GAME:

IDENTIFYING THE TOMATO

FROM SCIENCE TO POLITICS

"You say to-may-toe and I say to-mah-toe..."

The Gershwin Brothers weren't the first to argue about what to call this delicious fruit. In fact, calling it fruit, while botanically correct, was once a matter of legal dispute. How did the tomato name game get so complicated?

The words "tomato" or "tomate" are derived from the Aztec term for the plant, *tomatl*, and this has come to be the common way the inhabitants of much of Europe and the Americas refer to the fruit.

Lycopersicon is the name of the genus to which the tomato belongs. We tend to think of science as an exact art, but in the sixteenth century when the tomato was first brought to Europe, the naming of plants was in many ways as much a matter of romance as botany.

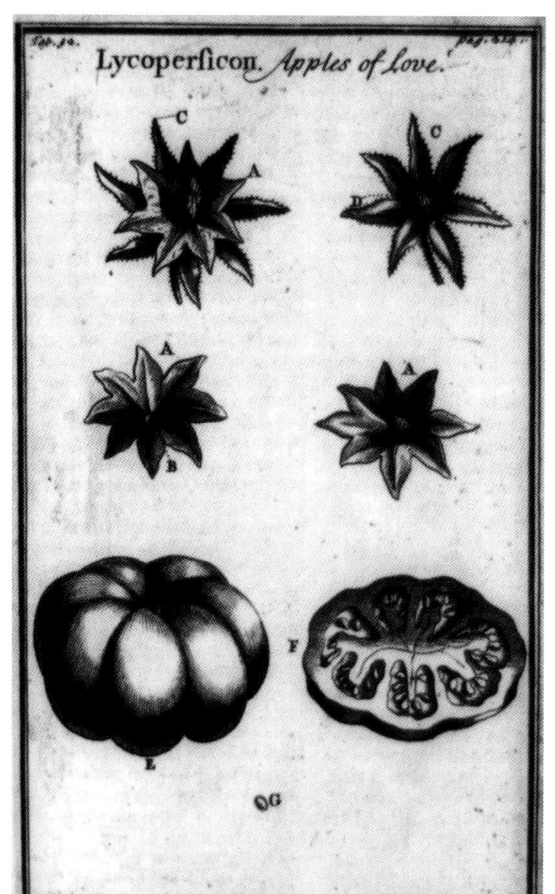

FROM WOLF PEACH TO GOLDEN APPLE

Lycopersion (note the slightly different spelling) was the Latin word that Galen, an ancient Greek physician who named many of the plants of his region, gave to the family of plants to which it was later determined the tomato belonged. In time the term was transformed to *lycopersicom* or *lycopersicon*, which means "wolf peach." It's a lovely term, and you might assume that the wild, slightly fuzzy, peach-hued skin of the tomato from Central America inspired it, except the name was around long before the tomato arrived in Europe, applied to plants that would later be determined its kin.

More romantic names for the tomato were to come,

however. The Italian *pomodoro* derives from pom d'or, or apple of gold, the common name people gave the tomato in that country. The first published reference to the tomato in Europe is from an Italian herbalist, Pietro Andrae Matthioli, who noted that tomatoes were green at first but ripened to a golden color. In a subsequent edition of his herbal in 1554, he noted that tomatoes were also red.

THE DARK SIDE OF THE STORY

Matthioli classified the tomato with mandrake plants, and here the tomato picked up a little character by association. The mandrake had been believed to be an aphrodisiac since at least Biblical times, making an appearance in Genesis as a love potion. So soon people began to refer to the tomato as a "love apple," and it was considered for some time to have special powers in this realm.

But not all the family associations were quite so pleasant. The eggplant is also in the mandrake family, *Solonaceae,* and herbalists of this era likely noted that the fruit, leaf structure, and flowers of the three plants are quite similar. Members of this family are commonly referred to as nightshades, and as the name suggests, they had a shady reputation. Although the fruits were consumed (Matthioli recommends cooking tomatoes as you would eggplant, in oil with salt and pepper), the plant itself can be toxic. It's possible that this explains in part why there was reluctance to embrace the tomato as a food item for some time in certain parts of Europe and North America.

MR. TOMATO GOES TO WASHINGTON

Eventually the tomato staked its claim on tables worldwide, however, and that sparked the next level of confusion in the name game. Botanically speaking, the tomato is clearly a fruit, in fact, a berry. But the Supreme Court of the United States decided to forego this little piece of evidence when setting a landmark (at least for tomatoes) decision in the late 1800s.

Prior to the American Civil War, tomatoes had been shipped from southern states to the north for sale in the weeks before they ripened in gardens there. The profits were considerable, and the demand for tomatoes didn't slack off when the war interrupted this trade. Savvy suppliers began to ship in tomatoes from the Bahamas and Bermuda. This trade continued after the war ended and soon many of the islands in the Caribbean were growing tomatoes and other vegetables for import to the United States.

American growers, concerned about the competition, convinced Congress to pass the Tariff Act of 1883. Under its jurisdiction, a 10 percent duty was levied on any imported vegetables.

Knowing his botanical rights, importer John Nix paid the duty on a load of West Indies tomatoes in the spring of 1886, but he did so under formal protest. Subsequently he filed suit against the collector saying the vegetable tax had been levied unfairly since the tomato was, in fact, a fruit. The case wound its way through the court system and ended up in the Supreme Court in 1893. Nix had his attorney read dictionary definitions of the tomato as fruit into the records and called several expert witnesses to testify as to its botanical classification.

But the court decided that the question was not whether the tomato was scientifically a fruit–a fact of which there was no dispute–but whether it functioned as a fruit or vegetable when it was regarded as a provision. (And, it should be noted, when it functioned as a taxable item, as well.) The decision of the court notes that in addition to tomatoes, cucumbers, squash, beans, and peas are also "fruit of the vine," yet like the tomato, all are used in the kitchen as vegetables and so would be classified (and therefore taxed) thusly.

Perhaps the court will have to reconsider its decision as more and more contemporary chefs find delightful ways to use the tomato in desserts and sauces that emphasize its fruitiness.

And perhaps this confusing Supreme Court decision helps to explain why in 1981 President Ronald Reagan's budget director, David

Stockman, decided it made perfect sense to declare ketchup a vegetable rather than a condiment. Stockman's declaration was made as a part of suggested budget cuts in the federally funded school lunch program. Stockman didn't fare as well as the Supreme Court, however. The proposal was killed and the idea publicly ridiculed for years.

Prior to this, Reagan's most memorable association with the tomato was his role in the 1942 film *Juke Girl* in which he played an itinerant tomato picker and left wing union organizer.

Botanically a fruit, legally a vegetable, the tomato performs well in either role in the recipes on the following pages.

Tomato, White Bean & Shrimp Salad

Serves 4

2 cups/473 ml cooked white beans (navy or cannelloni)

2 stalks celery, finely diced

2 carrots, finely diced

1 red bell pepper, finely diced

1 tablespoon olive oil

salt

white pepper

3 bay leaves

2 teaspoons paprika

16 large shrimp, shelled

2 tablespoons fresh minced basil

mixed greens

2 large tomatoes, sliced (use two varieties of different colors for a dramatic presentation)

- In a mixing bowl, combine the white beans, celery, carrots, red peppers, and olive oil. Toss and season with salt and white pepper to taste. (You can make this part in advance and chill, if you wish.)
- Half fill a medium saucepan with water. Add the bay leaves and paprika and bring to a lively simmer, uncovered. Add the shrimp and poach until the flesh turns pink and white. Remove from hot water and plunge into cold, then drain and allow to cool.
- When ready to serve, add the shrimp and minced basil and toss well to mix.
- Place greens on four salad plates and top with slices of tomato. Cover with white bean and shrimp mixture. Serve immediately.

A fantastic medley of flavor, color, and texture on a single salad plate.

Green Tomato Pie

Serves 6 to 8

1 recipe Perfect Pie Crust, page 22
1 cup/236 ml brown sugar
1/2 cup/118 ml white flour
pinch of salt
2 tablespoons butter
2 1/2 cups/590 ml green tomatoes, thinly sliced
1 medium-size, tart green apple, quartered, cored, and thinly sliced
1 tablespoon apple cider vinegar
1 egg
1 tablespoon water

- Preheat oven to 375°F/190.5°C.
- Roll out the pastry and place one crust in the bottom of a 9-inch/22 cm pie plate with crust evenly hanging over edge of the pan.
- Mix together the sugar, flour, and salt and sprinkle half on the bottom of the unbaked pie shell in the pan, then dot with the butter.
- Spread half the sliced tomatoes over the flour mixture and lay the apples over that. Top with the rest of the tomato slices.
- Sprinkle with the vinegar and spread the rest of the flour mixture over all.
- Top with the second pie pastry and make several vent holes with a fork. You can also cut the second pastry into even strips to weave a lattice crust. Crimp the edges to seal.
- Beat together egg and water, and brush on the top crust.
- Place in oven and turn to 350°F/176°C for 50 minutes.

Green tomatoes make this pie tangy and creamy at the same time.

Perfect Pie Crust

- *3 cups white flour*
- *2 teaspoons sugar*
- *1 teaspoon salt*
- *½ cup/118 ml vegetable shortening, chilled*
- *6 teaspoons butter, chilled*
- *½ cup/118 ml ice water*

- In a medium-size bowl blend the flour, sugar, and salt.
- Cut chilled shortening and butter into small pieces and add to flour. Using your fingers, two knives, or a pastry blender, cut the shortening and butter into the flour until the mixture looks like coarsely ground cornmeal.
- Sprinkle the water, 2 teaspoons at a time, over the mixture, using a fork to gently toss until the water is absorbed. You want a moist dough that holds together, not a soggy mass. You may not use all of the water, depending on how moist the flour is to begin.
- Pat dough together and divide into two equal balls. Shape each into a disk and cover with plastic wrap. Refrigerate for at least 4 hours.
- When you are ready to bake, remove the dough from the refrigerator and roll it out on a lightly floured surface with a few even strokes of a floured rolling pin. If the dough is difficult to roll at first, allow it to relax at room temperature for a couple of minutes. Roll each crust until approximately 10 inches/25 cm in diameter.

Hearty Cream of Tomato Soup

Serves 8

1 cup/236 ml chopped onion
1/2 cup/118 ml chopped celery
1/2 cup/118 ml chopped carrots
1 1/2 tablespoons olive oil
2 pounds/1 kg tomatoes, chopped
1/2 tablespoon minced fresh basil
1 1/2 tablespoons flour
1 1/2 cups/354 ml water
3/4 cups/177 ml cream (or creamy soy milk)
1/2 tablespoon honey
salt
pepper

- Sauté the onion, celery, and carrots in olive oil until very tender. Puree in a blender or food processor and return to pot.
- Mix in the tomatoes and basil, then sprinkle the flour over all. Add the water, stir to mix, then simmer on low heat until the tomatoes are very soft.
- Again puree the mixture and return to the pot.
- Stir in the cream (or creamy soy milk) and honey. Simmer until the soup thickens, then salt and pepper to taste.

Made with other garden vegetables, this tomato soup fortifies both body and soul.

"Here's what beats the heavy meats
The sugary cake and pie
It gives a zest to all the rest
And makes you strong and spry."
Menu
Campbell's
CONDENSED
TOMATO
SOUP
JOSEPH CAMPBELL COMPANY
CAMDEN, N.J., U.S.A.
"The biggest half"
What's the good of the best dinner without a good appetite to enjoy it?
Isn't a tempting delicious appetizer which makes the whole meal taste better and relish better really the "biggest" half of the menu any time?
Campbell's Tomato Soup with its captivating quality and flavor is not only a treat in itself but it makes all that goes with it doubly enjoyable and satisfying.
Made from the pure juice of solid, red, vine-ripened tomatoes blended with fine butter, granulated sugar and other choice ingredients, you never tasted anything more delicious. Have it for dinner or luncheon today.
21 kinds 15c a can
Campbell's SOUPS
LOOK FOR THE RED AND WHITE LABEL

Spicy Red Tomato Cake

Serves 8

2 cups/473 ml white flour
1 teaspoon baking powder
1/2 teaspoon ground nutmeg
1/2 teaspoon ground cinnamon
1/2 teaspoon ground cloves
1/2 cup/118 ml butter, softened
1 cup/236 ml sugar
2 eggs, room temperature
2/3 cups/157 ml tomato sauce
1/2 teaspoon vanilla extract
1/8 teaspoon salt

- Preheat oven to 350°F/176°C and prepare two 8-inch/120 cm cake pans with grease, flour or a lining of parchment paper.
- Sift together the flour, baking powder, nutmeg, cinnamon, and cloves.
- Use an electric mixer at medium speed to cream butter and sugar together until fluffy.
- Add the eggs one at a time and beat, scraping the sides of the bowl after each addition.
- Add tomato sauce, vanilla, and salt, and mix to blend.
- Add the flour mixture in four increments and blend well after each.
- Pour equally into prepared cake pans and bake at 350°F/176°C for 25-30 minutes, or until cake tests done.
- Allow to cool for 10 minutes in the pans, then turn layers out onto wire rack. Cool completely before frosting.

Cream Cheese Frosting

Enough to frost a two-layer cake

16 ounces/1/2 kg cream cheese, softened
8 tablespoons butter, softened
2 teaspoons vanilla extract
1 tablespoon lemon juice
2 pounds/1 kg confectioners sugar

- Using an electric mixer on medium speed, cream together the cream cheese and butter, stopping to scrape the sides of the bowl occasionally so the two ingredients blend evenly.
- Add vanilla and lemon juice and beat to blend in.
- Switch to low speed and gradually add the sugar, beating until all is incorporated and the frosting is fluffy.

Tomatoes add zip and color to this tempting spice cake.

FROM COZUMEL TO THE COSMOS:

THE TOMATO TRAVELS THE WORLD, AND THEN SOME

"A world without tomatoes is like a string quartet without violins." There are few who would disagree with these words from the late author, Laurie Colwin, writing in *Home Cooking*, a collection of her food essays from *Gourmet*. Yet there once was a time when such a world did exist.

Traces of the first tomato plants have been found in western South America. The coastal mountains of Ecuador, Peru, and parts of Chile still have wild tomato plants akin to these growing there. But there are no signs that the early people of South America cultivated these tomatoes or used them in culinary fashion. There is evidence, however, that natives of Central America—the Maya and other pre-Columbian peoples—began growing and using tomatoes as a food around the thirteenth century. For a major world food source, this is a fairly recent start-up date.

How the tomato migrated from South to Central America is not known for sure. It's assumed that tomatoes made their way from South America to the Galapagos Islands via the digestive systems of sea turtles; and it's likely that the plant migrated from South to Central America by biological locomotion as well. What is known is that the natives of the Yucatan cultivated the plant. The wild tomato was two-celled. At some point a genetic mutation produced a tomato with multiple cells. The Mayans (early genetic engineers?) selected seed to favor this mutation.

The Aztecs later adopted the plant and named it *xitomatl*, meaning large *tomatl*, because of its resemblance to the smaller, sour *tomatl* (or tomatillo), a very distant cousin. With the Spanish Conquest of 1519, the tomato began to show up in written accounts of the New World, but because of the similarities in name, it's often unclear if European writers were referring to the actual tomato or the tomatillo. Both made their way to Europe on return voyages, but it was only the tomato that caught on as a food there.

THE TOMATO ARMADA SAILS!

The Spanish proved to be more prolific disseminators than ocean-traveling turtles. Signs of cultivation in indigenous cultures in South America appear after the Conquistadors arrived there. They also distributed tomatoes in the Caribbean

IT'S ASSUMED THAT TOMATOES MADE THEIR WAY FROM SOUTH AMERICA TO THE GALAPAGOS ISLANDS VIA THE DIGESTIVE SYSTEMS OF SEA TURTLES

and in the Philippines, from whence it migrated first to Southeast Asia and then to Asia as a whole.

The Spanish took the tomato to Europe, and the tomato took a liking to the Mediterranean. By the beginning of the seventeenth century, tomatoes were being cultivated and used widely in recipes in both Italy and Spain. But northern Europeans did not immediately follow suit. (Ironically, in recent years some of the greatest stars of the heirloom tomato movement have been Russian, Czechoslovakian, Bulgarian, and German tomatoes.)

The English were wary of the fruit. They began to grow it for its ornamental value at the end of the sixteenth century and eventually used it for medicinal purposes, but the tomato was not regarded as a food item among the English for some 200 years after its introduction. Some accounts from that period suggested it could be poisonous to anyone living in a colder climate than that of the Mediterranean. Botanist John Gerard's *Herball*, first published in 1597 and highly regarded in its time, calls the tomato "the Love Apple." Gerard writes: "In Spaine and those hot regions they use to eat Apples (tomatoes) prepared and boiled with pepper, salt and oyle: but they yield very little nourishment to the body, and the same naught and corrupt." These days, however, the tomato is the fourth most popular fruit in the United Kingdom.

It was Portuguese slave traders who took the tomato to Africa, and it's likely that tomato seeds returned back across the Atlantic with slaves to the Americas.

The tomato also arrived in North America along the Florida coast with the Spanish, probably in the early seventeenth century, and spread northward as Spanish settlements did. Both British and French Huguenot settlers brought tomatoes to North America, as well; as did those who migrated to the southeastern coastal regions from the Caribbean. In 1774 the first published references to the tomato in North America appeared in two gardeners' calendars, from North and South Carolina. In each it was referred to as both vegetable and herb.

THESE DAYS FRESH TOMATOES FROM HOLLAND, CHILE, EVEN NEW ZEALAND CAN BE FLOWN IN AND APPEAR IN PRODUCE AISLES

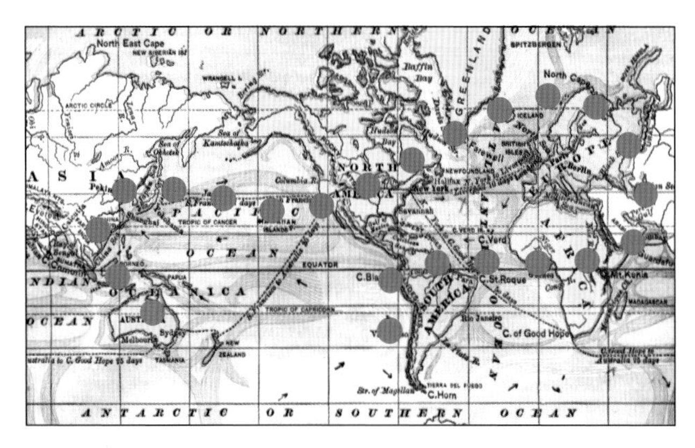

THE MAKING OF A CASH CROP

The tomato was initially more popular as a food in the southern states, although a liking for it spread to the northern parts of America where it was grown primarily for ornamental and medicinal purposes. By the early 1800s, however, it appeared regularly in recipes around the country, although sometimes a little strangely. In the classic *Miss Leslie's Complete Cookery: Directions for Cookery*, the most popular cookbook printed in North America in the nineteenth century, Philadelphian Eliza Leslie cautions that tomatoes "will not lose their raw taste in less than three hours cooking."

At about this time, articles on cultivating the tomato also became staples of North American farmer's almanacs and periodicals. Commercial seed producers began trying to develop strains which consistently produced fruit with smooth skin (the original tomato was deeply ribbed), good color, full flavor, and strong growing characteristics.

Up until the Civil War, commercial tomato production was dominated by southern growers. During the war, canned tomatoes became a staple of the Union Army and factories were built to produce them. Farmers in Delaware, Indiana,

BY THE EARLY 1800s, TOMATOES APPEARED REGULARLY IN RECIPES AROUND THE UNITED STATES.

Maryland, New Jersey, Ohio, and Pennsylvania flourished as they supplied product to the canneries. After the war, demand for canned vegetables increased throughout the country, and with canneries and production farms established, these regions became primary growers. (Today Florida is the largest producer of tomatoes for fresh sale in the United States while California holds the honors for tomatoes for processing.)

Fresh tomatoes began to be shipped by rail in the United States in the late 1800s, and by sea from the Bahamas and other islands in the Caribbean

from the Civil War on. These days fresh tomatoes from Holland, Chile, even New Zealand can be flown in and appear in produce aisles from Portland to, well, Portland. The United States remains the world's largest consumer and producer of tomatoes and tomato products. Italy is the second largest supplier of tomato products in the world, followed by Spain. Of late, several Latin American countries, Turkey, and China have been emerging as potentially strong tomato exporters as well. (The consumption of fresh tomatoes in China is tremendous compared to other countries: almost 40 pounds/18 kg of fresh tomatoes eaten per capita annually to about 17.5 pounds/8 kg for the United States. But there is almost no interest in processed tomatoes there compared to, say, the substantial 99 pounds/45 kg consumed by each person in the United States every year.)

HOMEGROWN STILL RULES

Commercial production has not slaked the consumer's desire for fresh tomatoes from the home garden, however, as the racks of seeds and flats of starts at garden outlets every spring attest, or the bins and baskets of fresh-picked vine ripe tomatoes at farm markets throughout the summer

CAZE

assure. The greatest interest appears to be in the realm of heirloom tomatoes. These old-fashioned varieties have been neglected by the commercial growing industry because they don't ship well; grow in shapes, colors and sizes that are irregular; are difficult to harvest or vulnerable to pests and disease. Home-growers and farm market customers swear, however, that when it comes to sheer pizzazz on the tongue, these older varieties can't be beat. (There is more on heirlooms on page 48. In the garden section on page 60, Barbara Ciletti describes how to get your essential tomato garden of both heirlooms and hybrids to grow.)

Tomatoes appear in gardens throughout the world, now, and in 1984 NASA sent tomato seeds into space–more than 12 million. After hanging out in the stratosphere for half a dozen years, the seeds were retrieved by the space shuttle Columbia and distributed to teachers to grow in their classrooms. Some people raised fears that the tomatoes might contain some strange outer-space mutation and so should not be eaten. But more intrepid souls ventured a taste or two as it became evident that the tomatoes differed in no significant way from their kin who had remained earthbound. Professor Robert McCullough at Ferris State University in Big Rapids, Michigan, was one of the recipients of the space seeds and in the summer of 2003 he harvested his thirteenth generation of tomatoes from space. As he has with preceding harvests, he ate some, canned some, and saved the seeds for the next year. You could say that with those classroom gardens and McCullough's ongoing harvest, the tomato has become a truly *universal* cultivar.

Around the world and back again, the traveling tomato found its way into countless regional dishes such as those on the following pages.

Lebanese Lamb, Okra & Tomato Stew

Serves 8

2 tablespoons bacon fat or butter
2 cups/473 ml onion, thinly sliced
2 pounds/1 kg lamb, cubed
2 cloves garlic, minced
1/4 teaspoon ground cumin
1/4 teaspoon ground ginger
1/4 teaspoon ground nutmeg
1/2 teaspoon ground cayenne
1/2 teaspoon ground coriander
4 cups/946 ml tomatoes, peeled, seeded, and chopped
1/2 cup/118 ml red wine
2 pounds/1 kg fresh okra, trimmed and sliced in 1/2-inch/1.25-cm lengths
3 cups/708 ml water
salt
pepper
juice of 1 lemon

- Melt 1 tablespoon of the bacon fat (or butter) in a heavy pot over medium heat. Sauté onion until translucent, then add the lamb and brown.
- Stir in the garlic and all the spices and simmer for 5 minutes, stirring occasionally so the mixture doesn't stick. Add the red wine and tomatoes and set aside.
- In a separate pan, melt the rest of the bacon fat (or butter) and sauté the okra for approximately 3 minutes, or until just golden, but not browned. Add the okra to the other ingredients.
- Add water, then salt and pepper to taste. Simmer 30 to 45 minutes over medium heat, until the lamb is tender. Stir occasionally to keep from sticking. If you are preparing in advance, refrigerate the stew at this point and warm up when you are ready to serve.
- Just before serving, adjust salt and pepper and stir in fresh lemon juice. Dish up immediately.

NOTE: If you are not serving all of the stew at once, add a spritz of fresh lemon juice to each bowl before serving—but not to the pot.

A spicy Middle Eastern stew tasty enough to inspire a party!

Muddle

Serves 12

1/2 pound/1/4 kg bacon, cut into 1/2 inch/1.25 cm pieces
2 leeks, cut into 1/2 inch/1.25 cm pieces
3 stalks celery, diced
2 red peppers, diced
1 teaspoon thyme
1 teaspoon tarragon
2 bay leaves
1/2 teaspoon cloves
1 1/2 teaspoon red pepper flakes
4 large tomatoes
zest of 1 orange
1/2 cup/118 ml tomato paste
2 pounds/1 kg potatoes, peeled and cubed
1/2 cup/118 ml chopped parsley
2 cups/473 ml shrimp stock
4 cups/946 ml water
2 pounds/1kg firm, white fish, cut into 1-inch/2.5 cm pieces
3 cups/708 ml peeled shrimp
salt
pepper
6 hard-boiled eggs, peeled

- Fry the bacon in a heavy soup pot until it begins to brown. Add the leeks, celery, and peppers, and sauté until the vegetables are tender.
- Using a spice grinder or mortar and pestle, crush thyme, tarragon, bay leaves, cloves, and pepper flakes into a powder. Add to pot and simmer about 5 minutes.
- Meanwhile, coarsely chop tomatoes and orange zest together in a blender. Add to soup pot along with tomato paste, potatoes, and parsley. Continue simmering for 10 minutes.
- Add shrimp stock, water, fish, and shrimp. Simmer a few more minutes until fish is cooked. Add salt and pepper to taste.
- The muddle is ready to serve now, but refrigerating it overnight will allow the flavors to mingle even more.
- When ready to serve, ladle into bowls and top each with half a hard-boiled egg.

Nothing could be finer than this classic fish stew from the Carolinas.

Lecso

Serves 6

1 cup/236 ml diced bacon
2 cups/473 ml diced onions
3 cups/591 ml coarsely chopped tomatoes
2 cups/473 ml peeled and cubed potatoes
1 green pepper, cut in strips
1/2 teaspoon ground turmeric
4 cloves garlic, minced
2 teaspoons paprika
water
salt
pepper
2 cups/473 ml chopped kale

- Sauté the diced bacon in a soup pot over medium heat until it starts to brown. Add the onions and cook until translucent.
- Add the tomatoes, potatoes, green pepper, turmeric, paprika, and garlic. Add enough water to just cover and simmer on medium heat until the potatoes are tender but not falling apart.
- Salt and pepper to taste, then add the chopped kale and simmer for 10 more minutes. Serve hot.

Hungry Hungarians have feasted on this traditional stew for centuries.

Gazpacho

Serves 8

- Peel and dice tomatoes. Puree with garlic, olive oil, salt, and cayenne.
- Mix in remaining ingredients except for melon and croutons. Refrigerate overnight.
- When you are ready to serve, add melon and garnish with croutons.

2 pounds/1 kg tomatoes
3 garlic cloves, crushed
4 teaspoons olive oil
1 teaspoon salt
1/2 teaspoon ground cayenne
2 teaspoons lemon juice
1/4 cup/59 ml red wine vinegar
2 celery stalks, diced
1 large red onion, diced
1 large cucumber, peeled, seeded, and diced
1 large green bell pepper, diced
3 cups/708 ml tomato juice
1/2 cup/118 ml parsley, chopped
1 cup/236 ml water
1 cup/236 ml diced cantaloupe and/or honeydew melon
croutons

The Spanish knew exactly what to do with tomatoes: Gazpacho!

Thai Salad of Papaya & Tomato

Serves 4

1 green papaya
2 tomatoes (sweeter varieties are best)
6 Thai chilies, seeded and finely chopped
2 stalks celery, diced
5 lime leaves, finely chopped
2 cups/473 ml green beans, blanched and cut into 1-inch/2.5 cm lengths
3 cloves garlic, finely chopped
juice of 1 lime
½ red onion, sliced
lettuce
Ginger Dressing (below)

- Peel the papaya, halve and remove seeds. Chop into ½-inch/1.25 cm cubes.
- Core tomatoes and chop into ½-inch/1.25 cm cubes.
- In a large mixing bowl combine all ingredients and toss with Ginger Dressing.
- Cover individual serving plates with washed lettuce leaves and place salad on top to serve. Or, if you prefer, toss lettuce with other ingredients and serve salad from a large bowl.

Tie into this snappy Thai salad that combines two fantastically flavorful fruits.

Ginger Dressing

Dresses 4 salads

1 teaspoon minced ginger
2 tablespoons apple cider vinegar
1 tablespoon sugar
¼ cup/59 ml peanut oil

- In a blender, combine ginger, vinegar, and sugar on medium speed. Slowly pour in oil until all is blended.

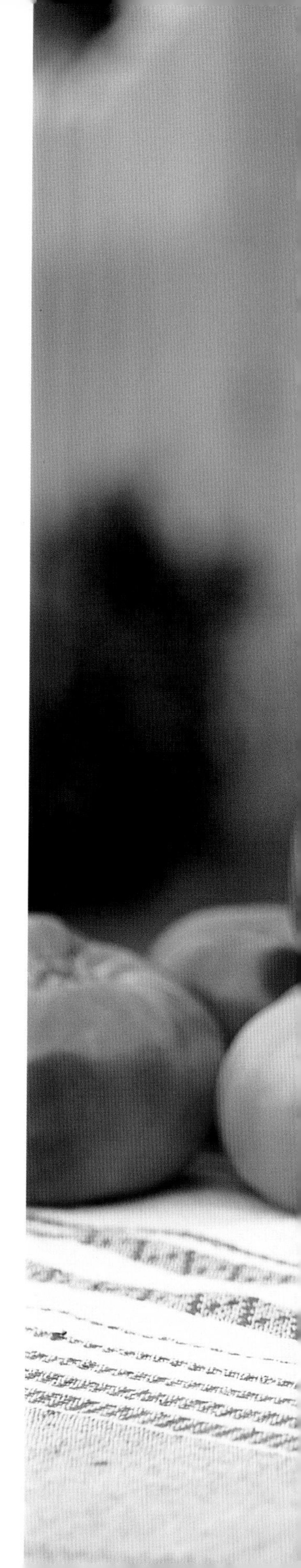

Malai Kofta

Serves 4

Kofta:

- *1 cup/236 ml cooked cauliflower florets*
- *2 large potatoes, peeled and cooked*
- *2 slices stale bread, diced*
- *1/2 tablespoon salt*
- *1 tablespoon flour*
- *1 tablespoon cornstarch*
- *1 teaspoon garam masala*
- *3/4 cups/177 ml feta or paneer cheese*
- *1/4 cup/59 ml bread crumbs*
- *1/2 cup/118 ml oil*

Sauce:

- *5 Roma or paste tomatoes, chopped*
- *3 medium onions, chopped*
- *1/2 tablespoon oil*
- *1 tablespoon minced ginger*
- *5 cloves of garlic, chopped*
- *1 teaspoon ground turmeric*
- *1/2 tablespoon salt*
- *1 teaspoon garam masala*
- *1 teaspoon ground chile*
- *1/4 teaspoon red pepper flakes*
- *1/2 tablespoon crushed cardamom*
- *1/2 cup/118 ml cream or coconut milk*
- *2 cups/473 ml water*

- *4 cups/946 ml cooked rice*

- In a bowl, mash together all of the kofta ingredients, except the oil and bread crumbs. Set aside for 20 minutes while you make the sauce.
- To make the sauce, sauté tomatoes and onions in oil until tender.
- Add ginger, garlic, turmeric, salt, garam masala, chile, red pepper flakes, and cardamom. Continue cooking for 5 minutes.
- Remove from heat and puree in blender or food processor until combined.
- Add cream or coconut milk, then add water until sauce is the consistency of a thin spaghetti sauce. Pour back into saucepan and keep warm over very low heat while you fry the kofta.
- Form the kofta mixture into round patties the size and shape of golf balls. Roll each in bread crumbs to coat.
- In a frying pan, heat the oil to a hot but not smoking temperature. Fry kofta, a few at a time and turning, until brown all over. Place on rack over paper towels to drain, then transfer to pot with sauce.
- To serve, place a cup of cooked rice in the middle of each plate, then ladle sauce with kofta over the rice. Serve immediately.

Note: Prepared garam masala may be found in Indian or gourmet groceries.

For a heavenly delight,
marry tomatoes with Indian spices and cream.

WHAT'S OLD IS NEW AGAIN:

THE HEIRLOOM GARDEN

They're on the covers of major food and garden magazines. They're making appearances on the Food Network and can be seen in all of the best restaurants. Experts are devoting whole books to them. Yet it's fair to say that many people, even those of us who seek them out at local farmers markets or plant a few in the yard, still don't know exactly what heirloom tomatoes are.

As their name suggests, heirloom tomatoes are old, as in, the seeds have been passed down for a some time. Heirlooms are open-pollinated varieties that have been grown for several generations. Exactly how many generations is not explicitly spelled out, but some have suggested that a plant should have been around for at least 50 years to be officially called an heirloom. On the other hand, open-pollinated varieties are currently being bred and introduced into the marketplace without such long pedigrees. Because of their exceptional qualities (particularly of taste), and because they are open-pollinated, which

means their seeds can be passed down so they can become heirlooms (as opposed to most hybrids), these newer varieties are often referred to as heirlooms shortly after their introduction.

Microbiology professor Carolyn Male, Ph.D., has been a primary force in the heirloom movement, having grown over 1,000 varieties of tomatoes and introducing numerous ones through Seed Savers Exchange, an organization devoted to preserving heirloom plants. Her book, *100 Heirlooms for the American Garden* is an in-depth guide for the devoted heirloom tomato grower. She and Craig Le Houllier, another respected tomato expert, have determined four distinct types of heirlooms.

The first is the commercial heirloom, which includes open-pollinated varieties introduced to the marketplace prior to 1940. The second is the family heirloom, passed down through generations (not specifically in the same family, by the way; close friends count). These weren't generally available through seed catalogs until quite recently. The third is a created heirloom that is formed by deliberate hybridizing to create a plant which is then de-hybridized through subsequent generations. (Seeds are saved from the plants and fruit that show the least deviation from the hybrid. They are planted, and subsequent seed saving occurs until no deviation shows up in the plant. At that point the plant is a new heirloom.) The final category is the mystery heirloom, the product of natural cross-pollination.

OPEN POLLINATED VS. HYBRID

And now seems like the appropriate time to explain the difference between an open-pollinated plant and a hybrid. A hybrid is the product of an arranged marriage, of sorts. The male flower of a plant of desired characteristics and of a pure variety is intentionally crossbred with a female of a different variety, also with desirable characteristics. The offspring of this arranged union is a plant that exhibits the best characteristics of each parent, often in a greater degree than either mom or dad.

Hybridizing makes it possible to breed out tendencies to disease, to increase yield, to extend the harvest season, and to create tomatoes of greater size, roundness, specific color or taste. Of late, hybrids have received a bad rap, but there is nothing intrinsically wrong with a hybrid, and as you can see from the description of the created heirloom, many an heirloom is initially the product of intentional hybridization. Hybrids have earned a bad reputation in part because much of contemporary breeding has concentrated on commercially desirable characteristics (firmer skin for ease in harvest, packing, and shipping; early production to make the most of market prices; uniform appearance to sell the tomato at the grocers), some say at the expense of flavor. But there are many hybrids that have been bred for characteristics which serious tomato lovers savor as well, and plenty of gardeners and eaters alike have favorite hybrids, such as the ever-popular Big Boy, Early Girl, and Celebrity.

But a serious flaw in the hybrid process is this: The seed is unstable, and succeeding generations planted from seed simply may not germinate. Those seeds that do germinate are not guaranteed to produce like the first generation hybrid. In fact, they are apt to produce fruit that is increasingly inferior to the first generation of the hybrid and to the parents. In other words, you have to buy new seed or starts each year to be sure of a consistent product. This is a desirable characteristic for most commercial seed companies and nurseries, and truly, the annual cost of tomato seed (or starter plants) is not an outlay that most home gardeners agonize over. But the downside of the popularity of hybridized tomatoes, which overwhelmingly dominated the latter half of the twentieth century, is that the varieties of tomato seed and plants available to most gardeners dramatically decreased to focus on the most popular, most reliable, and most conventional hybrid types.

Enter the heirloom renaissance, or the comeback of open-pollinated plants. Open-pollinated plants grow from the seed of the same plant annually. Growers can save seed from this year's crop, sow it the next year, and generally expect plants that produce fruit with the same characteristics and of the same or better quality than the parent the year before.

A SEED SAVING RENAISSANCE

Calling the current fascination with open-pollinated and heirloom varieties a "comeback" is a little misleading. Obviously people have been growing open-pollinated tomatoes since the beginning of tomato-time. Some families have passed down tomato seeds for generations. Gardeners and farmers have saved seed from their favorite plants and sown, swapped, and sold the seeds or starter plants from them forever. But, understandably, open-pollinated seed has not been the primary focus of most contemporary commercial seed companies. And in the latter half of the twentieth century, hybrid plants became increasingly the choice of

HIS CAREER AS AN HEIRLOOM SEED SAVER BEGAN WHEN HE BIT INTO A BRANDYWINE AND "THE SWEET JUICE WAS DRIPPING OFF MY ELBOWS BEFORE I HAD A CHANCE TO SWALLOW."

both commercial and home gardeners. As an overwhelming number of farmers and gardeners began to depend on commercial hybrids and fewer and fewer folks continued to save seed from open-pollinated plants, some varieties began to disappear altogether and others became the province of just a few dedicated, isolated planters.

The organic and back-to-the-land movements of the 1970s sparked renewed interest in the concept of seed saving. In 1975 a group of young North Americans committed to the concept of preserving heirloom vegetables and maintaining diversity in plant stock formed a cooperative called the Seed Savers Exchange. Headquartered in Decorah, Iowa, the nonprofit organization now has more than 8,000 members and lists 12,000 varieties of vegetables whose seed is available to them, including some 1,000 non-hybrid tomatoes. Similar seed-saving organizations now exist around the world in countries such as New Zealand, Switzerland, France, Great Britain, Australia, and Ireland.

By the 1980s some smaller seed companies began dealing exclusively in open-pollinated and heirloom seeds. (One of these, Southern Exposure Seed Exchange, near Thomas Jefferson's Monticello in Virginia, had the "seed" of its existence planted by a tomato. Founder Cricket Rakita has said his career as an heirloom seed saver began when he bit into a Brandywine and "the sweet juice was dripping off my elbows before I had a chance to swallow.")

THE HEIRLOOM GOES OUT TO DINNER

General interest in heirlooms was given a huge boost when top restaurateurs discovered that these old-fashioned varieties not only tasted exceptional but offered a panoply of color, texture, shape, and size perfect for creating a buzz about what was on the plate. Heirlooms started appearing on menus in surprising combinations. Creative chefs looking for new ways to feature the tomato to have been recently focusing on the fact that it is a fruit and making it a primary ingredient in desserts. (Tomato trend for the future? The return of freshly made tomato aspic featuring other heirloom vegetables and fruits in its crimson—or green—gel.)

By the first summer of the new millennium, a few swank restaurants in select places were featuring tomato tastings set up like wine tastings, with several varieties available to sample, crusty bread and olive oil for clearing the palate, and notepads to record flavor specifics and favorites. Of late, restaurants that focus on seasonal products have begun to feature fresh, locally grown, heirloom-dominated tomato samplers on their late summer menu.

BY THE FIRST SUMMER OF THE NEW MILLENNIUM, A FEW SWANK RESTAURANTS WERE FEATURING TOMATO TASTINGS.

Savvy diners soon wanted to get these outstanding tomatoes at home, and so increasingly heirlooms and old-fashioned varieties began to appear in local farm markets, and the seed catalogs specializing in them began to see a boom in business. Now even the major seed companies market heirloom seeds along with the ever-popular hybrids.

EVERY 'MATER TELLS A STORY, DON'T IT?

In addition to their vibrant sensual qualities, heirlooms offer the consumer another pleasure as well. Many of them have evocative names and some of them come with provocative pasts—or at least interesting stories connecting the planter to real people and real places across time and distance.

RIGHT: DR. CAROLYN'S YELLOW CHERRY TOMATOES

For instance, an actual Aunt Ruby was responsible for Aunt Ruby's German Green, one of the most popular of the ripe-when-green heirlooms. She lived in Greenville, Tennessee, and possessed seed that had been passed down in her family for generations. In 1992 she shared the seed with Bill Minkey of Darien, Wisconsin, who shared it with Seed Savers Exchange in 1993. When Ruby Arnold died in August of 1997, she left behind a legacy that continues to delight tomato aficionados around the world.

AUNT RUBY'S GERMAN GREEN TOMATOES

Then there's the geography of the Broad Ripple Yellow Currant, a small tomato that was found growing in a street crack at 56th and College in Indianapolis, Indiana, in the Broad Ripple Neighborhood. Or the global collegiality represented by the Druzba, a Bulgarian heirloom whose translated name means "friendship," and the Crnkovic Yugoslavian Pink, brought to this country by Yasha Crnkovic, who wanted to share a tomato from his homeland with his colleague, Dr. Carolyn Male. And chances are good that you can figure out who the variety called the Doctor Carolyn has been named after.

Then there is the metaphorical biology of the Black Sea Man, a strange, deep brown tomato that, when peeled, reveals veins beneath its skin which resemble a skeleton; or the literal image of the Oxheart, which proclaims its shape with its name. And what of the mythical resonance of the Stump of the World, suggesting a vast, primordial tomato, mother to them all, hidden in the earth's core?

Some tomatoes don't have stories but merely suggest them by evocative names. Did Nebraska Wedding spark a romance between growers? And what aches of longing and homesickness might the Hungarian Heart suggest, brought to the United States in 1901 by an immigrant from a small town outside Budapest.

Perhaps the most famously storied tomato is the much beloved Radiator Charlie's Mortgage Lifter. Its tale sounds apocryphal, but there was indeed a real Radiator Charlie and the folks at Southern

Exposure Seed Exchange did interview him in 1985. Their seed catalog carries excerpts from that conversation.

M.C. Byles, the real Charlie, was living in Logan, West Virginia, in the early 1930s when he decided to cross-breed four of the largest tomatoes he could then find. He planted one of the four in a circle formed by the others, then used a baby's ear syringe to cross pollinate the center plant. In subsequent years, he selected the best seedling from each lot to plant in the middle and repeated the process. He did this for six years until the variety was stable.

Radiator Charlie had earned his nickname because he had a radiator repair shop by the side of the road at the foot of a steep Appalachian hill where trucks were prone to overheat. In the 1940s he started selling his seedlings from his prized tomato plant for $1 each at his roadside shop. Soon gardeners were driving as far as 200 miles to get his tomatoes, and in six years Charlie paid off the $6,000 mortgage on his home from his sales.

Radiator Charlie's Mortgage Lifter tomatoes can grow so big, however, that some folks thought that was how they got their name—big enough to lift a house. The tomatoes average about 2.5 pounds (about 1 kg), but can grow as large as 4 pounds (2 kg).

The recipes coming up let you show off heirlooms by stuffing them with fabulous fillings.

STUMP OF THE WORLD TOMATO IN ITS GLORY

THE MORTGAGE LIFTER: BIG ENOUGH TO RAISE A HOUSE?

Heirloom tomatoes overflow with delectable savories.
What could be better on a summer day?

Tomatoes Stuffed Four Ways

- To prepare tomatoes for stuffing, first select large, firm ones with no blemishes—one for each person you are serving. (A tomato with a wide, flatter bottom sits better on a plate.)
- Slice off the top quarter of each tomato. Gently scoop out pulp and seeds, leaving a solid wall of tomato flesh around the perimeter of the fruit.
- Sprinkle lightly with salt and invert on racks for 30 minutes before righting and stuffing with any of the fillings here.

Grilled Fresh Tuna Stuffing

Fills 4 tomatoes

1 pound/$^1/_2$ kg fresh tuna
2 tablespoons olive oil
salt
lemon pepper
1 tablespoon parsley, chopped
1 tablespoon fresh chives, sliced thinly
2 cloves garlic, chopped
zest of 1 lemon
juice of 1 lemon
1 tablespoon fresh sweet basil, chopped

- It's easiest to cook the tuna if it's in three or four equally sized pieces. Coat each side with oil, then season with salt and lemon pepper. Place in a hot, nonstick skillet or on a hot grill and cook for about 3 minutes on each side, until the tuna is medium rare.
- Cool enough to handle and gently break into smaller pieces in a mixing bowl. Add all the other ingredients and mix gently.

Avocado Bulgur Stuffing

Fills 4 tomatoes

1 cup/236 ml cooked bulgur
2 tablespoons olive oil
1 avocado, peeled and diced
2 jalapeno peppers, minced
1 tablespoon diced red onion
1 tablespoon chopped fresh cilantro
1 small tomato, peeled and diced
juice of 1 lime
salt
pepper

- Combine first eight ingredients in mixing bowl, then season with salt and pepper to taste.

Eggplant Stuffing

Fills 4 tomatoes

medium eggplant
olive oil
1/2 cup/118 ml roasted red pepper
1 teaspoon crushed red pepper
1 tablespoon tahini
2 teaspoons minced garlic
juice of fresh lemon
salt

- Split eggplant in half lengthwise, rub with oil and cook under broiler or on grill until flesh is softened and skin is lightly charred. Set aside to cool, and when you can handle it, scoop out flesh.
- In a blender, combine all ingredients except the lemon juice. When everything is pureed, begin to add the lemon juice just to make it smooth. You want a firm substance, not a runny one.
- Add salt to taste. Refrigerate for 1 hour before stuffing tomatoes.

Crab Stuffing

Fills 4 tomatoes

1 pound/1/2 kg white crabmeat
2 tablespoons thinly sliced scallions
1 tablespoon capers, drained
2 teaspoons minced fresh parsley
1/4 cup olive oil
1 teaspoon Dijon-style mustard
salt
white pepper
4 lemon wedges

- Pick over the crabmeat to remove any bits of shell, then place in a bowl.
- Combine the crabmeat with the scallions, capers, and parsley, being careful not to break up any lumps.
- Whisk the olive oil with the mustard and season to taste with salt and white pepper. Toss with the crabmeat mixture, then adjust seasonings.
- Serve with lemon wedges on the side.

NOTE: Four hardboiled eggs, peeled and coarsely chopped, may be substituted for the crabmeat.

Herbed Tomato Aspic

Serves 12

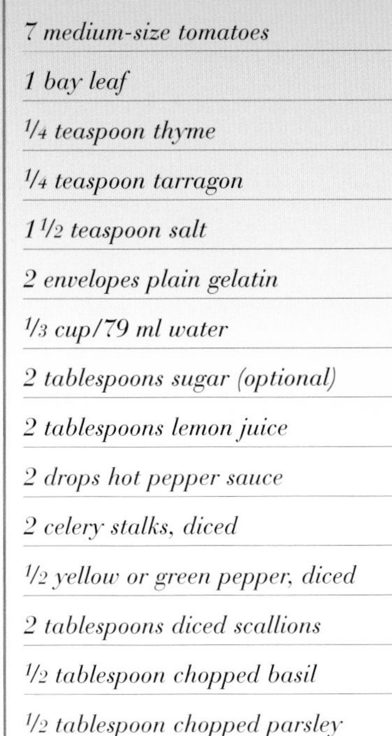

- *7 medium-size tomatoes*
- *1 bay leaf*
- *1/4 teaspoon thyme*
- *1/4 teaspoon tarragon*
- *1 1/2 teaspoon salt*
- *2 envelopes plain gelatin*
- *1/3 cup/79 ml water*
- *2 tablespoons sugar (optional)*
- *2 tablespoons lemon juice*
- *2 drops hot pepper sauce*
- *2 celery stalks, diced*
- *1/2 yellow or green pepper, diced*
- *2 tablespoons diced scallions*
- *1/2 tablespoon chopped basil*
- *1/2 tablespoon chopped parsley*

- Prepare a bundt pan or gelatin mold for the aspic by spraying lightly with nonstick culinary spray or oil.
- Remove the tomato stems and puree tomatoes. Bring to a boil in a saucepan with the bay leaf, thyme, tarragon, and salt, and simmer for 20 minutes, stirring often. Strain through a food mill to remove skins and seeds. Return to saucepan over low heat.
- Soften the gelatin in cold water for 3 to 5 minutes. Stir the gelatin, sugar (if using), lemon juice, and pepper sauce into tomato mixture, and continue stirring over low heat until the mixture is smooth. When it is smooth, remove the mixture from heat.
- Allow the mixture to cool until the aspic begins to thicken. (You can cool it at room temperature, over ice or in the refrigerator, but keep an eye on it so you know when it starts to thicken, and don't let it get too firm before the next step.)
- Stir in the celery, pepper, scallions, basil, and parsley. Pour the mixture into the prepared bundt pan or gelatin mold and chill for at least 3 hours.
- When you are ready to serve, prepare a platter with lettuce leaves. Run a knife gently around the edges of the aspic, then turn it over on the plate to unmold. Serve immediately, or refrigerate again until you are ready to serve.

Essence of tomato suspends the flavors of an herb garden in aspic.

THE ESSENTIAL TOMATO GARDEN PRIMER

by Barbara Ciletti

"That potato, he grows in the dark or in the damp cellar with his pale lank roots; he has no flavor; he lives underground. But the tomato, he grows in sunshine; he has fine, rosy color, an exquisite flavor; he is wholesome, and when he is put in the soup you relish him."

MICHEL FELICE CORNE

The nineteenth-century Italian immigrant artist Michel Corne understood that the allure of the tomato lies not only in the kitchen, but also in the garden. Corne is sometimes erroneously credited with eating the first tomato in America, changing attitudes and turning it from an ornamental plant to a culinary one. In fact, tomatoes had been eaten for quite some time before Corne arrived in the United States, but in the early years of its world-wide dissemination, the tomato was prized as much as an object of delight in the garden as it was on the table. That romance continues today as, despite its availability in the marketplace, the tomato continues to be the most popular home garden plant in the world.

Growing tomatoes is not difficult and can be done in almost in any situation, even on a fire escape in the heart of the city. For beginner and veteran alike, successfully raising tomatoes means paying attention to a few details that can make all the difference. Knowledge of the essentials can ensure not one but several crops of ongoing produce from late spring until frost.

TOMATO TALK

With a few key language tools in place, you can plan a tomato patch specific to your kitchen and cultivate a crop that yields great produce for the table, the canner, or for snacks. Here are some terms you need to know.

DETERMINATE: These plants grow to a certain height, then blossom and produce all of their fruit in 14 to 21 days. Since determinate plants yield their produce all at once, within a limited period, they are good if you plan to do a lot of canning of either tomatoes or sauce, or if you want to fill the freezer with an abundance. When doing your garden planning, bear in mind that determinate plants usually require a cage and infrequently require a stake or trellis, depending upon the variety. These plants are bushy, not viny, and are good for smaller gardens or containers used in places where space is a consideration.

INDETERMINATE: These plants grow more like vines than bushes and will continually blossom and yield fruit until the first frost. These are the plants you want if you would like a regular supply of fresh tomatoes all season. Some indeterminates provide huge fruit and bountiful yields. Others offer smaller yields of perhaps two to four tomatoes at a time. Some varieties will attain a height of 3 feet (about 1 meter) or more. Others are truly giants, reaching for the sky, upwards of 6 feet (about 2 meters). Indeterminates twist, turn, sprawl, and without training or a trellis or two, will take over your garden.

Beefsteak tomato: Large and slightly elliptical with deep red flesh and fewer seeds than other varieties. Other slicing tomatoes exist, but these are stunning for sandwiches and platters.

Cherry tomato: About 1 inch (2.5 cm) in diameter and available in red, orange, yellow, or ripe green varieties, the cherry tomato usually arrives early in the garden, and the plants that bear it have the ability to keep producing well into late summer. These tomatoes are great for snacks, salads, shish kabobs (with or without meat), or any occasion that requires a splash of color and a mouthful of flavor, with no cooking necessary.

Globe or slicing tomato: Usually mid- to late-season tomatoes with a round and fairly regular shape that makes them ideal for slicing. These are also good in salads.

Green tomato: Several varieties of tomato are ripe when green (Aunt Ruby's German Green and Zebra Stripe, for instance), but the culinary definition of this term indicates a tomato that has reached full size but has not yet ripened. Green tomatoes are a classic main course or side dish when sliced, breaded, and fried. You can also broil or bake them in pies or casseroles, or cook them in soups, or salsa.

Heirloom and hybrid tomato: For a full discussion of heirloom and hybrid distinctions, see page 48.

Ground cherry: In the past this was a term for all husk tomatoes but recently it has come to refer specifically to those which have a very sweet taste and are the size of a cherry or smaller. They have been used in jams and sauces for centuries, but there has been some question as to whether they are edible raw. Recently, however, they have been showing up on restaurant menus both fresh and cooked and being devoured both ways with no ill effect.

Husk tomato: Fruit of the genus Physalis, the most common being the tomatillo. Although they resemble small ones, these are not actually tomatoes. They are characterized by the papery calyx which surrounds the fruit.

Maturity: This is the number of days from transplanting seedlings to the first mature fruit. Early season plants take 55 to 68 days, midseason plants take 69 to 79 days, and late-season plants take 80 days or more to maturity.

Paste tomato: Though you can use these for salads and slicing, they are the best for sauce, paste, and cooked salsa. If you want to can or sun dry a full crop, seek out a hearty determinate variety that provides big yields of consistently sized produce.

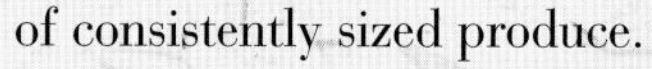

Husk tomatoes are not actually tomatoes.

PEAR TOMATO: The size of a cherry tomato and used the same way, but shaped like a tiny pear or teardrop.

PLUM TOMATO: Also called a Roma, this paste tomato is an oblong shape.

POTATO-LEAF: Tomato foliage with no indentation on the leaf margin.

PURPLE TOMATO: Skin colors range from a violet-tinged deep pink to purple so deep it appears dark brown or almost black. Flesh may range from red to brown with hints of purple.

REGULAR-LEAF: Tomato foliage with indentations on the leaf margin.

STRIPED TOMATO: Any tomato with striations of a different color, such as Zebra Stripe (green with yellow) or Candy Stripe (yellow with red).

WIDE ADAPTATION: Tomato varieties that have adapted to cultivation in several zones possess this characteristic. (Always read your seed packets and refer to a zone chart to select varieties that thrive in your area.)

ZONED OUT

Once you've succumbed to the grower's tomato passion, take time to become acquainted with the varieties and how they match the specific soil and climate conditions that define your garden patch. Not surprisingly, these South American natives can not only take the heat but thrive in it. Yet, a number of tomato varieties are also well suited to cooler climates and shorter growing seasons. Some even do well in areas where the long, hot days have little promise of moisture.

Seed packets and catalogs almost always feature a zone map. Paying attention to what grows well in your zone will help you make a first, common sense decision about what can or cannot thrive in your particular patch of dirt. Your county agricultural extension office can offer plenty of region-specific information; and a nearby university can also be a valuable resource. Once you know the limits and possibilities of your space, choose tomato varieties accordingly. If you wish to produce tomatoes through the entire growing season, a blend of early, mid-, and late-season varieties should be part of your garden plan.

TO SEED OR NOT TO SEED?

Shopping by way of seed catalogs offers a broad range of tomato opportunities. You can salivate over cherry tomatoes, heirlooms, exotics, paste tomatoes, beefsteak, and other varieties as you plan your dream garden. You can imagine a color palette that nearly spans the spectrum from deep purple to bold crimson to tantalizing orange to buttercup yellow to lime green to almost white.

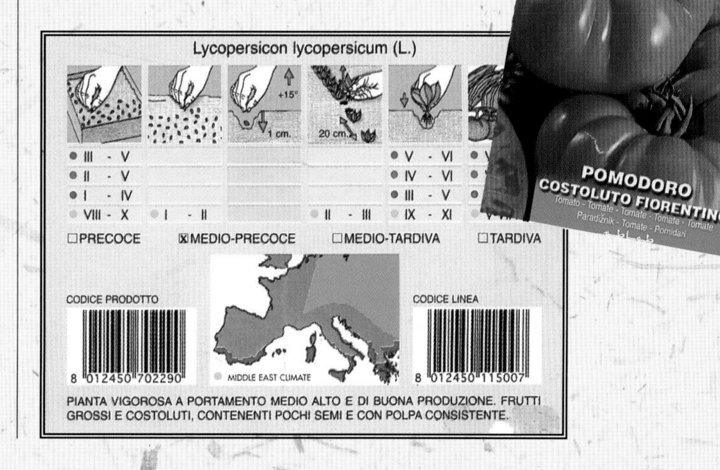

The tastes available to you range from citrus tart to peachy sweet; the sizes, from that of grapes to those which require two hands to heft. Starting from seed allows you to choose from a much wider array since there are many varieties that can't be purchased as starter plants. Most young plants for sale have been grown because they are compatible with the greatest span of growing conditions. That broadens the marketability of the plants but limits the choices for you.

Yet, there are advantages to buying starter plants. Many folks who want to grow their own tomatoes don't have the time or space for coaxing plants from seed indoors. Starter plants allow you to get a jump-start in the garden since commercially started plants are often ready to go before home seedlings have come into their own. And with the growing interest in heirlooms and unusual varieties these days, it's possible to find some older, rarer starter tomato plants at farm markets or quality nurseries.

Savvy gardeners will often do both: start seeds indoors, and later, as the outdoor growing season nears, purchase plants to fill out the roster. No matter where they begin their indoor life, your plants will need to gradually acclimate to life outdoors before you transplant them in the garden. (More on this later.)

Regardless of their start, tomatoes universally require the same nutrients, air, soil, and environmental conditions to thrive.

SOW AND GROW!

You should begin germinating seeds six to ten weeks prior to the estimated transplant time. When would that time be? After the final frost of the season and when the ground is warm. Of course, the date of the last frost in a specific region can vary from year to year, but you can estimate when it's apt to be by referring to the weather patterns of your region in the past. If you're not sure, ask another grower or a local nursery when the normal tomato transplanting time for your area occurs. (In the bluegrass and more northern regions of Kentucky, for instance, folks like to get their tomato plants out by the day after Derby Day, which is the first Saturday in May. That would seem late for folks who live closer to the equator but early to tomato growers in chillier northern latitudes.)

You can use small peat pots or flats for sowing the seed. Make sure the containers are filled about two-thirds of the way up with a light, sterile germination mix (not plain dirt) to give the seeds that extra push to sprout. You can purchase a prepared mix or you can make your own from one third vermiculite, one third perlite, and one third milled sphagnum moss. The medium should feel spongy and the finer the texture the better. Sow seed 1/2 inch/1 cm deep, in the center if you are using a peat pot, or 1 inch/2.5 cm apart in flats.

Until the seeds germinate, you need to keep the soil (not air) at a temperature between 70°F to 90°F/21°C to 32°C, with 85°F/29°C as the ideal. Try to keep the temperature consistent, day and night. Fluctuations send mixed signals during the germination process, and cooler conditions often impede or slow down the transformation from seed to sprout. You can purchase a heat mat

from a garden supply store to maintain a steady soil temperature. Many growers swear by setting their flats on top of the refrigerator, which always seems to be warm but I think this is unwise since the chill when you open the door could be considerable. There are soil temperature gauges if you want to monitor precisely.

Keep the germinating mix moist but not wet. Too much water can settle in, and an overall damp condition encourages disease. Germination times vary according to the variety, but you should have sprouts within six to ten days.

The new seedlings require a sunny window or grow light for stability and continued development. Remember that while mature tomato plants love the heat, seedlings can't take direct sunlight. They grow best in a place that gets a maximum of indirect sun throughout the day.

You will also need to begin feeding your seedlings with a weak compost tea solution at this point, since the sterile growing medium supplies no nutrients. Don't overdo it, however. Patience heralds the day here. Let nature take its course. Your job is to provide a balance of nutrients, moisture, heat, and light. You nurture the seedling, and nature will reward you with big and tasty tomatoes to enjoy.

If your seedlings are in flats, you can transplant them to larger pots, preferably 2 inches (5 cm) wide from seven to ten days after they have germinated. Before transplanting, gently remove all leaves from the stem, with the exception of those roughly 1 inch/2.5 cm from the top. You should end up with something that looks like a tiny tree, with a slender trunk enhanced by foliage at its uppermost end. The lack of stem foliage encourages additional root growth by removing what would otherwise be competition for nutrients that flow from the soil to the plant.

Fertilize your seedlings with fish emulsion by following the instructions on the package. At this point, transfer your young plants to a sunny spot, but one where they can develop at a cooler temperature (60°F/15°C) for another 14 to 21 days. While the warm start is ideal for germination and root development, the gradual reduction in temperature spurs more foliage development and initiates the inevitable transition to life in the garden. The vigorous foliage will ultimately yield an abundance of blossoms, which of course should yield an abundance of tomatoes. In fact, when plants are about three weeks old, organic gardening authority Shep Ogden, co-owner of The Cook's Garden seed and supply company in Burlington, Vermont, recommends lowering temperatures to 55°F/13°C indoors at night, with a slight rise to 65°F/18°C indoors during the day. This process should be continued for about 12 days before resuming the standard 65°F/18°C day and night. Ogden says this encourages more flowers on the plant, and hence, more tomatoes.

You will need to transfer your plants one more time, to 4-inch/10-cm pots, before they finally take up residence in the garden. Once again, remove leaves, this time any you find up to 2 inches/5 cm from the top of the plant. Tuck the cleared stem beneath the soil. Water the pots consistently, yet with a light touch. You'll want enough water to keep the grow mix from drying out but not enough to get things really wet.

Your plants are ready to settle permanently into the great outdoors once any danger of late frost has passed. Before this point, you should be preparing to make the big tomato move. For two weeks prior to the time you intend to transplant your tomatoes into the garden, take them outside for an increasing amount of time each day. Begin by leaving them out in the morning for two to three hours, then extend the time through the afternoon for the next two to three days, then leave them out from morning through supper time for two to three days. Keep the plants in a warm yet sheltered spot that gets even, gentle sun. When your plants have adjusted to the full range of temperatures and light, your tomatoes are ready to mature, blossom, and yield in the garden.

ABOUT THE HEAD START CLASS

If you decide to purchase plants in addition to or instead of sowing seeds, buy early in the season and look for plants already nestled into their 4-inch /10-cm pot lifestyle. At this point the plants should be about as wide as they are high, with healthy green foliage. Buying plants early in the season allows you to pick from the best selection and to control growth conditions as soon as possible.

As you select plants at the local nursery, beware of those bearing small fruit. Many beginning gardeners purchase plants with tiny tomatoes on them, thinking that will put them a step ahead in the maturation process. Not true! If anything, plants that set fruit while still indoors have a more difficult time adjusting to the outdoors. Why? The process of transplanting to the garden provides a mixed message to the tomato seedling with fruit. The whole purpose of transitioning to the garden is to encourage the growth of foliage, which then leads to the development of blossoms. The young plant with fruit has already skipped this process. The result is a confused adolescence in which the plant doesn't know whether to spend its energy producing more foliage or maturing the fruit, and so it does neither very well. Once again, the lesson here is not to rush the process. The best-tasting tomatoes come from plants that have been allowed to develop vigorous foliage and acclimate to cooler temperatures before setting fruit.

Buy small, compact plants, not tall, leggy specimens. Tomatoes in 4-inch /10 cm pots should be at the development stage that is focused on producing healthy roots, not the foliage. So, if you see tallish young plants, with plenty of top foliage, pass on them.

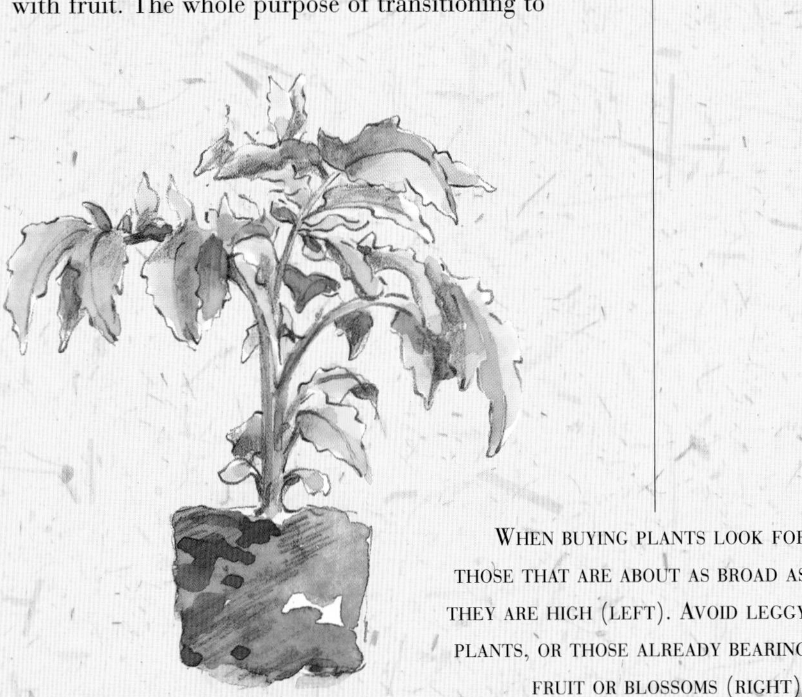

When buying plants look for those that are about as broad as they are high (left). Avoid leggy plants, or those already bearing fruit or blossoms (right).

Buy smaller plants that don't look so mature, and keep reminding yourself that it was the tortoise who won the race, not the hare.

Check out the tomato pot. Turn it on its side to make sure that the plant isn't pot-bound, with roots trailing up the outside. Pot-bound plants are stressed-out plants.

TIPS FOR A VIBRANT LIFE OUTDOORS

Tomatoes prefer loose soil with a balance of nitrogen, potassium, and phosphorus. Particularly note that too much nitrogen can reduce plant yields. The pH range for the soil should be from 5.8 to 7.0.

Tomatoes are lusty eaters, so keep the soil full of nutrients. They like water, organic compost, and fertilizer. A hearty supply of nutrients will help

create lots of foliage to protect the stems and will lead to plenty of blossoms and, ultimately, tasty fruit.

The preferred daily soil temperature for tomatoes outside is 70°F/21°C. Mulch can help maintain that temperature even if the weather doesn't cooperate.

While plants are growing, plenty of water at the ground level is required, but never water a tomato plant from the top down. If you water from the top of the plant, healthy foliage will keep much of the water from hitting the ground and soaking through to the roots, which is where the tomato needs it. In addition, leaves with water on them are more susceptible to scorching in the sun. Another way to make sure that moisture goes into the dirt where it is needed is to keep leaves up off the ground. Soggy leaves lying on the floor of the garden simply promote disease.

All tomatoes, even those cultivated in cooler climates and regions with shorter days, love the sun. Nothing is better than full daytime sun for the photosynthesis and the maturation of your tomato crop.

Tomatoes also need room to grow, although determinate and indeterminate plants have different space requirements. Consider the assumed mature size of your plants (along with characteristics such as bushy plants versus vines) as you dream about your tomato patch. You'll need to plant smaller bush plants approximately 15 to 20 inches/38 to 50 cm apart; determinate, or larger bush type tomatoes should be in rows 24 to 30 inches/60 to 75 cm apart, and vines that promise to extend to a height of 4 to 10 feet/1 to 3 m should be planted 36 to 45 inches/90 to 115 cm apart.

AND THEY'RE OFF!

Tomatoes, like good racehorses, need a little training and some judicious coddling. Once transplants have been set in the garden, apply a ground cover, such as black or red plastic, to keep the earth warm. This will help protect against plant and root shock until the growing season is in full force and temperature fluctuations aren't a daily gardening factor. If you choose, you can elect to use a floating row cover, which also keeps the ground warm and protects young plants against wind, pesky flying insects, and the neighborhood cat (maybe!). A floating row cover is a lightweight black fabric that is permeable, so water can reach the soil underneath. You use it to cover the entire planting bed; simply cut holes in it where you will place your plants.

Tomato plants are either determinate or indeterminate, and that helps determine how you train them in the garden. Bushy determinate tomatoes thrive when surrounded by a cage that keeps bottom leaves off the ground and provides plenty of air circulation throughout the plant, even on breezeless summer days. As the plants grow up inside the cage, branches will twine around its metal supports, and eventually the supports will hold the fruit off the ground as it matures, providing it with more sunlight and protecting it from pests and disease.

Indeterminate plants generate vines that can grow longer than 10 feet/3 m. For that reason, stakes or a trellis are the order of the day. You can set up a trellis similar to one used for pole peas and beans. There are a variety of different stringing methods you can use, but the object is

to run string from the bottom to the top of each plant, across the top of all plants in the trellis. Keep the string slack, leaving room to adjust it to meet the needs of growing and maturing vines. The string encourages the tomato vine to grow from a single stem with leafy branches that lead to fuller branches with flowers.

Pruning axial stems encourages the plant to grow up the trellis, not out. (An axial stem is one that shoots off perpendicularly from the main stem, leading the plant to grow in a horizontal direction.) Pruning foliage at this point also promotes better air circulation and thereby fortifies against plant disease. Once the plants climb the trellis and reach the top, pinch off the growing point. The plant will stop growing, the nutrients will head for the fruit and your tomatoes will begin to plump up and gain their intended mature color.

GENTLE STROKES FROM GENTLE FOLKS

Once tomatoes stake their claim in the garden, they thrive under the care of the sensitive gardener. Like most creatures, great and small, tomatoes have their likes and dislikes. Here are some things you can do to please yours.

FERTILIZE: Plan to feed your tomatoes every two to three weeks with fish emulsion, compost teas, or the natural fertilizer of your choice.

STROKE: Brushing lightly with your hand two to three times a week will promote stocky, vigorous bush determinates. By brushing the vines upward, you can promote their ongoing spread as they set their sights on the sun. Brushing tomato plants encourages the development of a hormone called cytokin, which promotes strong, thick plant stems. Nature strokes tomato plants by way of breezes through the foliage. When your hands make a similar, gentle motion it's more of a good thing.

MULCH: Plastic, grass clippings, and straw are all good mulches for tomatoes. Red or black plastic will keep the soil warm. Black plastic protects against annoying weeds, while red plastic doesn't offer as much weed protection. What red plastic may do, however, is reflect sunlight up to the tomato, and thereby encourage the maturation of fruit. Disciples of red plastic mulch boast of earlier harvest and bigger fruit. You may leave plastic mulch on throughout the season, but many growers prefer to use different types. When the soil is consistently warm, for instance, you can switch to grass clippings or straw, which do a better job of keeping warm soil moist. (Cold water and cold soil in the tomato garden invite disease and pests.)

PRUNE: This procedure is optional for determinate, bush-type tomatoes, but it's a must for vine

REMOVE THE SUCKERS (NON-FLOWERING STEMS) THAT GROW IN BETWEEN THE MOTHER STEM AND THE LEAF CROTCHES.

or indeterminate tomatoes. Without a bit of training and pruning, these plants will grow in all directions at once. Once indeterminates are in the ground for a week or so, you can begin the process of removing the suckers (non-flowering stems) that grow in between the mother stem and the leaf crotches. (Wait for two leaves to develop first, then remove the third leaf that grows between them. See illustration on page 72.) Doing so keeps plant growth focused toward the main stem. The main stem gets fortified, produces and sets blossoms, and you get tomatoes. Otherwise you'll get The Vine That Ate the Garden, and perhaps the rest of the neighborhood.

FRIENDS AND NEIGHBORS

Just as soil conditions, sunlight, and warm temperatures affect the development and maturation of your tomatoes, so does the presence, or absence, of certain other plants. If you plan to cultivate a tomato crop among other vegetables and herbs, you will want to keep some things in mind.

Crop rotation is essential for a healthy garden. Planting the same plant in the same space year after year depletes the soil of the specific nutrients that plant needs. Planting different plants each year means different nutrients will be used and some that were depleted the year before may even be replenished. So tomatoes should not follow tomatoes in the same soil in consecutive years, but they also should not follow potatoes, peppers, or eggplant. All of these cultivars are members of the same family and therefore require the same soil nutrients.

In addition, there are some plants that tomatoes do not get along with in close quarters, and others with which they form a mutual admiration society, enhancing one another horticulturally. Tomatoes should not be planted among pole beans, corn, dill, fennel, or potatoes. These plants either sap the soil of nutrients essential to the tomato, or attract pests and disease that are harmful. Tomatoes can be planted with asparagus. In fact, they will help protect those tender stalks against the beetle that often plagues asparagus. Tomatoes also protect gooseberries against insects, and roses against black spot disease.

Your tomatoes will thrive when planted amid garlic. They also do well with basil, bush beans, carrots, celery, chive, cucumbers, peppers, head lettuce, marigold, mint, nasturtium, onions, and parsley.

GARDEN ENEMIES

Tomatoes are susceptible to several common pests and diseases. Nutrient-rich soil and healthy plants are your best guardians against both. Certain varieties are resistant to certain problems; and you'll find that information in the seed catalog or on the seed packet. Smart companion planting and crop rotation can also help, and sprays and applications used by organic gardeners, as well as the chemical kinds are also available. Your local agricultural extension office can provide information on what has worked well in your area in the past. Here are a few of the most common culprits and how they will appear in your garden.

You may think your tomato has fleas if you notice the tiny *black hopping beetles* that like to nibble on your tomato leaves, but most often what you will see are the little round holes they leave behind. *Blister beetles* don't do much damage to the tomato but can be unpleasant for the gardener who accidentally crushes one and discovers how they got their name.

Tomato hornworms (left) are also rarely visible until their damage is done. That's because these caterpillars are green like the tomato foliage and not-quite-mature fruit they love to consume. Handpicking them off the plants is a good way to control an invasion early in the season.

Nematodes are the trolls of tomato-dom, living underground and assaulting the roots of your plants. Crop rotation is the first line of defense in controlling nematodes. Look for resistant plants if you have been bothered with them in the past.

The fungus among us include *fusarium* and *verticillium*, both hearty, long-lived soil fungi; the first preferring warm earth, the second liking cold soil. Crop rotation and planting resistant varieties are the best ways to controlling both.

Alternaria is another fungal disease. This one preys upon weak or bruised fruit, causing decay around the stem and skin. These dark brown chains of spore cells are transmitted through the air and a part of regular outdoor life. A few spores on the foliage won't keep you from having a healthy crop. Heavily affected plants, however, can find relief from a spray of compost tea.

Tobacco mosaic virus (TMV) is fairly easy to control by keeping tobacco smoke and smokers, whose unwashed hands can spread the disease, out of the tomato patch. Once it enters your garden soil it can hang around for up to two years. The effect? Your tomatoes will be bitter, and you will reap smaller crops. If you are a smoker, you should know that most commercial growers require their workers who smoke to wash with a solution of bleach and water before handling plants. If your tomatoes do become infected with TMV (leaves will appear deformed and mottled), pull up the plants immediately and throw them away. Do not compost! The disease is tenacious.

Other problems may plague your tomatoes as well. Tomatoes can develop *skin cracks* when moisture isn't consistently available. If your garden climate provides many dry days, compensate with a watering system that will maintain consistent moisture. If not, your tomatoes will develop a tight skin from lack of water. Then when moisture does return and plump the inner flesh, the skin will rupture.

Blossom end rot frequently results from periods of dry weather and soil, which create calcium deficiency uptake. Once the plant is deprived of this nutrient, the fruit gets soft on the bottom and then turns leathery. At this point, you must remove the affected fruit. Blossom end rot does occur as frequently with trellised or caged tomatoes. Again, steady moisture will protect, and good drainage is essential.

Cat-facing is the term used for poorly formed or scarred fruit that has endured cool or very wet weather during the pollination period. Fortunately cat-faced tomatoes may not look pretty, but they taste just fine.

THE APARTMENT DWELLING TOMATO

Almost everyone longs to grow a tomato plant or two, even those of us without a postage stamp of lawn to call our own. Fear not! While the immense personality and flavor of the tomato can hardly be contained, its roots and foliage can, allowing it to thrive in large pots and containers on the smallest of balconies and porches. Just make sure the pots provide good drainage. When placed in sheltered areas that receive consistent sun and protection, your tomato plant will yield an admirable crop.

Large crocks that are 12 inches/30 cm deep and at least that in diameter make good homes for tomatoes. Small determinate tomato plants and varieties of small indeterminates that can be staked will yield handsomely in such homes.

Check your container garden at least once a day, and if the top inch of soil feels dry, water until it drains from the bottom of the container. Don't be afraid to water twice a day, if your tomato needs it.

Like their cousins in the garden, container tomatoes need mulch. Shredded leaves, grass, or peat moss all are attractive for plants that serve a decorative purpose as well as a culinary one, and each encourages water retention. Fertilize and prune your patio or porch tomato plant just as you would those in the garden patch.

Even the most urbane tomato perched on a big city rooftop can get attacked by weeds. Cultivate the soil to make sure it is loose and aerated, and pull any weeds as soon as they appear.

SO LITTLE TIME!

The best tasting, sweetest tomato is the tomato that stays on the vine, soaking up the sun, until it has reached the glowing pinnacle of its intended mature color. As the season progresses and your plants, laden with fruit, stretch in splendor, prune back the larger leaves, allowing additional sun and air to reach the fruit.

Once the tomatoes are ready to harvest, gently twist (don't pull!) the fruit from the vine. Allow tomatoes to rest at room temperature until ready to use. Refrigeration quells the flavor and changes the texture of your plump ripe fruit from succulent to mealy.

If you have tomatoes on plants as frost approaches, you can either cover the plants with light blankets or harvest the green produce and allow it to ripen indoors. If placed in paper bags or nested in newspaper, your green tomato will turn red. It won't be as vibrant as one that has ripened on the vine, but you'll get a good tasting tomato, nevertheless.

When that last tomato from the garden is gone, you may think wistfully of the varieties you didn't grow this year. You may marvel that there are so many more tomatoes to be cultivated and enjoyed. That's the time to set your sights on the next year, when you begin the process again. For once you've coaxed a seed into a plant, nestled your baby into warm soil, nurtured and protected it through the season, and plucked the reward warm and ripe from your very own vine, your passion for such pleasure will never end.

Following the chart of tomato varieties are a dozen recipes to help you make the most of your garden treasures.

THE COMPLEAT CHART FOR CULTIVATING & COOKING TOMATOES

Early Season

Variety	Garden Tips	Determinate	Indeterminate
Bush Early Girl	sturdy, hearty bushes that are highly productive	❖	
Early Girl	add mulch for good drainage and weed control; may need a cage		❖
Eva Purple Ball	German heirloom with high yields		❖
Fourth of July	vines produce plentiful harvest early and all summer long		❖
Juliet	high yields from bushes of miniature plum tomatoes		❖
Matina	heirloom from Germany; needs good drainage		❖
Santa Hybrid	prolific vines up to 7 inches/18 cm tall	❖	
Sugar Snack Hybrid	prolific vines start early and keep producing into cold weather		❖
Stupice	highly tolerant to cold and very productive		❖
Sun Gold	tall plants bear long clusters of fruit		❖
Sweet Baby Girl	compact, manageable plants		❖

Pest/Disease Resistance	Maturation and Fruit Size	Best Use
good; VF hybrid	fruit with big, full tomato flavor 8 to 10 ounces/227 to 284 g after 65 days	salads, slice, burgers
good; reliable in any climate VFF hybrid	bountiful harvest of solid, 4- to 6-ounce/113- to 170-g fruit after 52 days	snacks, platters, slice, stuff
good	small dark pink tomatoes are ready to harvest after 70 days	snacks, platters, slice
good	plentiful harvest of 4-ounce/113 g fruit is ready to pick after 49 days	slice, salads, snacks
good	very juicy, sweet cherry tomato variety that is ready to harvest after 60 days	snacks, salads, shish kabobs
good	small 2- to 4-ounce/57- to 113-g fruit with intense tomato flavor ripens after 58 days	snacks, salad, salsa
good	pear-shaped tomatoes, very sweet after 60 days	snacks, salads, shish kabobs
strong against pests	cherry tomatoes hang in full clusters and are ready to harvest after 65 days; known as the sweetest cherry tomato in the patch	snacks and salads
good	small early tomatoes that sparkle with flavor are ready to harvest after 52 days	slice, salads, snacks
good	fruity yet sweet clusters are ready to harvest after 57 days	snacks and salads
strong against tobacco mosaic virus	full clusters of compact, dark red fruit are ready to harvest after 65 days	snacks and salads

Mid Season

Variety	Garden Tips	Determinate	Indeterminate
Better Boy	hearty vines produce large crops		❖
Black Krim	Russian heirloom is prone to cracking but is a heavy producer; fruits set well even in hot weather		❖
Brandywine	this hybrid is an heirloom with lots of foilage and high yields		❖
Bush Big Boy Hybrid	compact and ideal for small gardens, cages, and tubs	❖	
Celebrity Hybrid	strong vines and good ground cover; 1984 All-America selections winner	❖	
Costoluto Genovese	Italian heirloom that produces well in hot weather and continues to thrive as cold weather approaches		❖
Floralina	tall, viny plants have ongoing, profuse production		❖
Heatwave Hybrid	perfect for hot climates where the daytime temperatures reach 90-96°F/32-36°C consistently	❖	
Sioux	yields a large harvest even in hot weather		❖
Rose	European hybrid with vines that produce bountiful clusters of fruit throughout the season		❖
Tangella	medium-size plants produce early in the season and continue to thrive into cold weather		❖
Sweet Tangerine Hybrid	bushy, relatively compact plants yield heavy crops early in the season but also produce well in hot weather	❖	
Valencia	this hybrid from Maine yields a crop that ripens earlier than most, a good choice for short growing seasons		❖
Wisconsin 55	a new 2003 hybrid that thrives in a variety of growing seasons		❖

Pest/Disease Resistance	Maturation and Fruit Size	Best Use
hearty VFN hybrid	big, red, smooth tomatoes are ready to harvest after 75-80 days	grill, slice, saute, stuff
good	large, dark red fruit 10-12 ounces/284-341 g with rich sweet flavor; ready to harvest in 75-90 days	slice, saute, stuff, grill
good	medium 8- to 10-ounce fruit is ready for harvest after 78 days	salsa, juice
strong disease resistance	large 10-ounce/284 g tomatoes, very sweet, juicy with excellent aroma are ready to harvest after 71 days	slice, platters, salads, salsas
strong VFFNTA hybrid	firm 7- to 8-ounce/198- to 227-g fruit ready to harvest in 70 days	slice, saute, sandwiches
good	large-ribbed, juicy tomatoes are ready to harvest in 78 days	salads, salsa
good	clusters of small (but not tiny) pear tomatoes, sweet and plump, are ready to harvest after 78 days	snacks, salads, shish kabobs
strong	plants yield uniform, round 7-ounce/198-g fruit after 68 days	salads, slice
good	big clusters of medium, round red fruit with intense, complex flavor are ready after 70 days	sandwiches, burgers, saute, salads
good	clusters of medium 8- to 10-ounce/227- to 284-g rosy red fruit are ready to harvest after 75 days; flavorful and sweet with a spark of tartness	salsas, salads, platters, slice
good	bright orange fruit in full clusters, slightly larger than the standard cherry tomato; fruit is sweet yet tart with a kiss of lemon	salads, snacks, platters
good	deep yellow-orange fruit is meaty, rich in flavor, and ready to pick after 68 days	shish kabob, saute, burgers, platters, stuff, roasting
good	large orange 8- to 10-ounce/227- to 284-g tomatoes are ready to pick in 76-80 days	slice, chop, saute, toss with pasta
good	large 8- to 10-ounce/227- to 284-g fruit is smooth, sweet, and abundant after 78 days	slice, chop, use raw or cooked

Late Season

Variety	Garden Tips	Determinate	Indeterminate
Ace 55	a hearty bush that bears a less acidic fruit	❖	
Aker's West Virginia	a treasured West Virginia heirloom that produces all summer long		❖
Amish Paste	an Amish heirloom that is a consistent producer of meaty fruit		❖
Big Rainbow	a beefsteak bicolor heirloom that produces consistently throughout the season		❖
Big Tomato Heirloom	tall viny plants produce huge red and yellow tomatoes.		❖
Box Car Willie	an excellent producer and a good choice for those who want to cultivate a single variety in one crop		❖
Burpee Supersteak	vigorous producer of exceptional, large fruit with very few cracks and blossom end marks		❖
Druzba	Bulgarian heirloom with profuse harvests of smooth, glossy fruit		❖
German Red Strawberry	tall viny plants yield oxheart tomatoes that look like big red strawberries		❖
Giant Pink Belgium	huge, hearty plants with good ongoing yield throughout summer and until cold weather		❖
Mortgage Lifter	a hearty heirloom that produces huge pink fruit		❖
Olena Ukranian	a vigorous Ukraine heirloom with good ongoing production		❖
Quarter Century Heirloom	short, sturdy heirloom bushes with crinkled leaves, good for container planting	❖	
Red Rose	a hearty cross between the Rutgers and Brandywine varieties		❖
Russian	heirloom oxheart with high productivity		❖

Pest/Disease Resistance	Maturation and Fruit Size	Best Use
strong against pests and wilt VF	medium-sized tomatoes are ready for harvest after 80 days. This tangy tomato with a pleasant, sweet dimension is good for those who prefer a low-acid fruit	salads, sandwiches, grill
good	large 16-ounce/453-g fruit with smooth skin, few cracks, ready to harvest after 85 days	salsa, salad, stuff, grill
good	oblong and meaty oxheart tomatoes are ready to pick after 85 days	pastes, salsa, saute, grill, roast
good	huge tomatoes, frequently more than 24 ounces/680 g, are sweet and ready to pick after 90-100 days	good for pan sauces, canning, sauté, grill, roast, cooked salsa
good	huge sweet and mild heirloom up to 22 ounces/ 627 g are ready to harvest after 85 days	tomato platters, slice for sandwiches and burgers
good	6- to 8-ounce/170- to 227-g tomatoes with intense flavor are ready after 80 days	sandwiches, salads, saute, snacks
strong against pests and disease	big and hearty, 16-24 ounces/453-680 g; provides a full beefsteak flavor with a touch more tartness after 80 days	stuff, roast, grill, sauce, canning
good	dark red and juicy 8- to 10-ounce/227- to 284-g tomatoes are ready to harvest after 80 days	salsa, sauces, juice, sandwiches, burgers
good	tart, meaty tomatoes of 10-14 ounces/284-397 g are ready to harvest after 85 days	saute, can, roast, grill, slice
good	tomatoes that average 24 ounces/680g, one of the sweetest varieties; ready to pick after 90 days	juice, sauce, wine, paste, stuff.grill
good	huge heirloom beefsteak up to 4 pounds/2 kg full of meat and ready to harvest after 85 days	can, sauce, paste, grill, stuff, roast
good	huge pink tomato, at least 16 ounces/453 g, with solid blemish-free flesh, are ready to harvest after 85 days	slice, stuff, roasting
strong disease-resistant plants perform well even in hot weather	medium red beefsteak tomatoes, 4-6 ounces/113-170 g, are ready after 85 days	slice platters, stuff, grill,
strong disease and crack-resistance	dense yields of 4- to 8-ounce/113- to 227-g meaty pink fruit are ready for harvest after 85 days	slice, platters, salad
good	heart shaped, meaty and intensely flavored fruit with solid meat is ready after 90-95 days	slice, salsa, sauce, grill, stuff

Late Season

Variety	Garden Tips	Determinate	Indeterminate
Sandul Moldavan	heirloom variety from the Moldavan region of Europe, produces vigorously throughout the summer and into cold weather		❖
Santa Clara Canner	vigorous vines produce large ongoing harvests		❖
Yellow Brandywine	tall vines with potato leaf foliage produce large fruit throughout the season		❖
Zogola	heirloom red beefsteak from Poland		❖

Paste Varieties

Variety	Garden Tips	Determinate	Indeterminate
Banana Legs	bushy prolific plants yield clusters of yellow fruit	❖	
Jersey Devil	viny plants with prolific yields		❖
Opalka	vigorous heirloom from Poland with dense, wispy foliage		❖
RomaVF	sturdy bushes that require weekly watering	❖	
San Marzano	vines that yield clusters of pear-shaped fruit		❖
San Marzano Redorta	Tuscan heirloom plum tomato with ongoing yield		❖
Viva Italia Hybrid	bushy, sturdy plants with full yields	❖	

Exotic Varieties

Variety	Garden Tips	Determinate	Indeterminate
Black From Tula	Russian heirloom that prodcues all season and continues even as the weather begins to turn		❖
Black Plum	tall vines produce a steady crop		❖
Cherokee Purple	a Tennessee heirloom, very productive all season		❖
Green Zebra	vines that produce a unique green eating tomato		❖
Persimmon	hearty vines that produce consistently		❖

Pest/Disease Resistance	Maturation and Fruit Size	Best Use
good	large 12- to 16-ounce/341- to 453-g tomatoes with sweet tart complexity are ready after 90 days	salsa, slice, grill, salads, can
good	large 10-ounce/284-g tomatoes with a good balance of meat and juice are ready after 80 days	slice, salads, sacue, salsa, can
good	tomatoes with a rich, smooth flavor, most more than 12 ounces/341 g, are ready to pick after 90 days	sauce, paste, grill, roast, stuff
good	large, juicy sweet tomatoes are huge, usually more than 4 inches/10 cm in diameter and ready to harvest after 85 days	slice, salsa, sauce, grill
good disease resistance	tapered 4-inch-/10-cm-long fruit with meaty yellow flesh ready after 75 days	great yellow tomato sauce and puree
good disease resistance	tapered banana-shaped fruit, 4-6 inches/ 10-15 cm long, is ready after 80 days	paste, sauce, salsa, grill
good disease resistance	tapered banana-shaped fruit, 4-6 inches/ 10-15 cm long, is ready after 80 days	good for paste, sauce, and snacks
high disease resistance	clusters of solid 3-ounce/85-g fruit are ready to harvest after 76 days	paste, sun dry, grill
good disease resistance	fruit 3 to 3 1/2-inches/7.5 to 9-cm long is meaty, flavorful and ready after 80 days	paste, sauce, puree, snacks
good disease resistance	large tomatoes are ready after 78 days	sauce, paste, snacks
high disease resistance	clusters of solid 3-ounce/85 g fruit are ready to harvest after 80 days	paste, sun dry, grill
good disease resistance	large 8- to 12-ounce/227- to 341-g deep reddish-brown, sweet fruit is ready to pick after 75 days	Platters, slice, salsa
good disease resistance	small oval tomatoes with plenty of flavor, ready after 82 days	snacks, shish kabobs
good disease resistance	intensely flavored, rich 10- to 12-ounce/284- to 341-g fruit is ready for harvest after 80 days	not a keeper
good disease resistance	medium 3-ounce/85-g green fruit is ready to harvest after 75 days	best fresh, salsa
good disease resistance	large, orange fruit, full of tomato flavor is ready to harvest after 80 days	salsa, sauces, salads, snacks

Tomato Sandwiches—A Picnic's Best Friends

THERE ONCE WAS AN EARL
'CROSS THE SEA
WHO DEVOURED TOMATOES WITH GLEE.

HE ATE THEM WITH SALT,
ONIONS, BASIL, AND MALT

BUT TWIXT SLICES OF BREAD—HOW YUMMY!!

EXTRA
VIRGIN
OLIVE

Tomato Stuffed Vegetarian Muffaletta Sandwich

Makes 8 sandwiches

1/4 cup/59 ml chopped carrots
1/4 cup/59 ml chopped onion
1/4 cup/59 ml chopped celery
1 cup/236 ml olive oil
3 cloves garlic
1 cup/236 ml green olives with pine nuts
1 cup/236 ml ripe black olives
1/2 cup/118 ml roasted red peppers
2 tablespoons white wine vinegar
2 tablespoons chopped parsley
8 hoagie sandwich buns
1 large purple tomato, sliced
1 large yellow tomato, sliced
1 large red tomato, sliced
1/2 pound/1/4 kg sliced provolone cheese
1/2 pound/1/4 kg sliced mozzarella cheese
3 tablespoons chopped fresh basil

- To make the olive salad for the muffalettas, sauté the chopped carrots, onion, and celery in a small amount of the olive oil until tender. Add the garlic and cook for 2 more minutes. Transfer to a food processor.
- Add the green olives, black olives, red peppers, white wine vinegar, and the remaining olive oil. Pulse ingredients to make a coarse but well blended mixture. Stir in the parsley.
- Slice the hoagie buns in half and scoop out a well in the bottom half of each. Layer the three kinds of tomato in the well and cover with sliced provolone and mozzarella. Top with the basil leaves; then drizzle with some of the olive salad juices.
- Spread a generous amount of olive salad on the top half of each sandwich and put the top and bottom together. Refrigerate overnight. When you are ready to eat, slice into wedges and serve.

Ratatouille Hoagie

Makes 8 sandwiches

- *1 medium-size eggplant*
- *1 pound/½ kg pattypan squash*
- *1 red onion, diced*
- *2 bell peppers, diced*
- *2 tablespoons olive oil*
- *1 cup/236 ml sliced mushrooms*
- *4 cloves garlic, minced*
- *1 pound/½ kg tomatoes, peeled, seeded, and coarsely chopped*
- *¼ cup/59 ml chopped fresh basil*
- *1 tablespoon fresh thyme leaves*
- *salt*
- *pepper*
- *8 crusty hoagie buns*
- *8 slices Swiss cheese*
- *alfalfa sprouts*
- *1 recipe for Basil Mayonnaise (below)*

- Preheat oven to 350°F/176°C.
- Peel the eggplant and cut into ½-inch/1.25-cm cubes. Cut the pattypan squash into ½-inch/1.25-cm cubes also.
- In a large pan that can go from stove-top to oven, sauté the onions and peppers in the olive oil until they begin to soften a bit but are not fully cooked.
- Add the squash, eggplant, and mushrooms and cook until they just start to become tender.
- Add the garlic and tomatoes and cook for 1 minute. Add the basil and thyme. Salt and pepper to taste.
- Cover the pan and bake for 30 minutes at 350°F/176°C.
- Split hoagie buns in half and scoop out a well in the bottom half of each. Fill with several heaping spoonfuls of ratatouille. (If the ratatouille filling has too much liquid, strain it before filling the buns.)
- Cover the filling with a slice of Swiss cheese and garnish with sprouts. Liberally coat inside of the top half of each bun with basil mayonnaise, top the sandwich and serve.

NOTE: If you want to make a vegan version of this sandwich, eliminate the cheese or use soy cheese. Substitute the Eggplant Filling for Stuffed Tomatoes on page 58 for the Basil Mayonnaise.

Basil Mayonnaise

- *¼ cup/59 ml basil leaves*
- *1 cup/236 ml mayonnaise*
- *1 dash hot pepper sauce*

- Using a blender, food processor, or mortar and pestle, macerate the basil leaves to form a paste. Blend with mayonnaise and hot pepper sauce.

Tomato Sandwich l'Oignon

Makes 8 sandwiches

1 or 2 large ripe tomatoes, peeled
1 large sweet onion
3 tablespoons mayonnaise
1 tablespoon mustard
16 slices of bread
salt
pepper

- Cut tomatoes into eight slices, 1/4 inch-/1.25 cm-thick. Do the same with the onion, then layer tomato and onion slices in a shallow dish. Cover and chill for at least 8 hours.
- When you are ready to make the sandwiches, stir together mayonnaise and mustard, and spread on one side of each bread slice. Discard the onion and use the tomato slices to make each sandwich, sprinkling lightly with the salt and pepper.

Tomato Sandwich
with Blackberry Chipotle Mayonnaise

Makes 4 sandwiches

- In a blender, puree the blackberries and pepper. Combine with the mayonnaise.
- Fry the bacon until crisp and drain on a rack. Leave about 1 tablespoon of bacon drippings in the frying pan.
- Slice red onion thinly and sauté in bacon grease over low heat until golden brown.
- Spread blackberry chipotle mayonnaise on one side of each piece of bread and assemble sandwich with bacon, onions, tomato, and lettuce.

1 cup/236 ml fresh blackberries
1 chipotle pepper
1 cup/236 ml mayonnaise
16 bacon strips
1 medium red onion
8 slices of bread
1 large tomato, sliced
lettuce

BLFGT

Serves 1

3 fried green tomato slices (recipe page 140)
4 bacon strips
3 lettuce leaves
mayonnaise
2 whole wheat toast slices

- Put them all together.
- Say yum.

Summertime is sandwich time with tomatoes picked right off the vine.

Grilled Pizza

Serves 4 to 6

Toppings:

12 plum tomatoes, sliced
4 Japanese eggplants, sliced
1 cup/236 ml chopped red onion
½ cup/118 ml chopped green bell pepper
5 cloves garlic, minced
¼ cup/59 ml olive oil
¼ cup/59 ml chopped fresh basil
1 tablespoon fresh thyme leaves
Dough for 4 personal (10-inch/25-cm) pizzas
3 cups/708 ml grated mozzarella
2 cups/473 ml grated parmesan

- In a bowl, mix the tomatoes, eggplant, red onion, bell pepper, garlic, and oil, and set aside. In a cup, mix together the basil and thyme.
- Preheat the grill to medium low. (Use some mesquite chips if desired.)
- When the grill is hot, roll out the dough into 10-inch/25-cm rounds. Brush one side with the olive oil and place on the grill, oil side down. Brush the top with oil. Grill until the bottom is golden brown, approximately 4 to 5 minutes.
- Turn over dough and immediately cover with the vegetable toppings. Sprinkle the herbs over that and top with the grated cheeses. Close the lid of the grill and bake another 4 to 5 minutes, until the cheese melts.
- Serve immediately.

Pizza is so fine hot off the grill
with tomatoes fresh from the garden

Sungold, Watermelon & Feta Salad

Serves 4 to 6

1 pound/½ kg Sungold or other golden cherry tomatoes
4 cups diced and seeded watermelon cubes
½ pound/¼ kg watercress
½ pound/¼ kg feta cheese, crumbled
salt

- Place first four ingredients in bowl and toss lightly to mix. Salt to taste. Serve immediately.

Zesty Tomato Garden Dressing

Makes about 2 cups

- In a bowl, combine all ingredients with a whisk. Because this dressing is a temporary emulsion, you must whisk or shake it before each use.

1 tablespoon fresh orange juice

1 teaspoon minced orange zest

1 tomato, peeled, seeded, and diced

1 tablespoon Dijon-style mustard

1 tablespoon cider vinegar

2/3 cup/157 ml minced scallion

1/2 teaspoon celery seed

2 tablespoons peanut oil

1/4 cup/59 ml olive oil

1 drop of hot pepper sauce

1/2 teaspoon salt

1/4 teaspoon freshly ground black pepper

Chock full of garden goodness, this tasty dressing is almost a salad in itself.

OH YES YOU CAN CAN, FREEZE, AND DRY

When you are planning your garden in February it may seem as if there's no possible way you could grow enough fresh tomatoes to satisfy your winter cravings. But if all goes well, in even the most modest tomato beds, by late August, you and all your kin will be satiated. What to do with the rest of the abundance? Depending on where you live, it is time to either "put up" or "put by" the surplus of your crop.

Both phrases mean the same thing: to prepare produce in some way to keep it palatable through the winter. Canning has been the home-grower's time-honored way to deal with tomatoes since James L. Mason patented glass jars with porcelain-lined screw tops in 1858. The ease of freezing has won over many a cook with ample freezer space in recent years, however. And drying, perhaps the oldest method of storing tomatoes, has made a robust comeback, as sun-dried tomatoes have become a popular ingredient in a multitude of dishes. Here's what you need to know about all of these.

A JARRING CONCEPT

You can put up tomatoes in jars using either of two methods. The first, and the one used most often, is the hot water bath method. The second is the pressure canning method. Both of these require glass jars made for the purpose of canning (an ordinary jar you have recycled from, say, peanut butter or mayonnaise may not hold up in either the hot water bath or pressure canner). Jars should be clean, and it is a very good idea to sterilize them in a boiling water bath for 10 minutes. Just before you are ready to fill them, remove the jars from hot water, turn them upside down on a clean rack, and allow them to dry for one minute.

You will need enough canning lids and rings for each jar. You can reuse the metal rings, but you will need unused lids to assure a good seal. Lids and rings should be boiled for 10 minutes and can be left in the hot water until you are ready to put them on the jar.

Low acid foods need to have acid added to them to prevent spoilage. Most tomatoes are right on the cusp of the safety mark when it comes to acid, although more and more low acid tomatoes are being bred. Short of doing a pH test in your kitchen, it is really not possible to know if your particular tomatoes are high enough in acid to forego the addition. So it is necessary to add lemon juice or citric acid to the tomatoes before processing. Add 1 tablespoon of lemon juice or 1/4 teaspoon citric acid per pint, and double that for quarts. Add the acid directly to the jars before adding the tomatoes. It will not significantly alter the taste. The only exception to this rule are unripe green tomatoes: they are more acidic than ripe ones and can be canned without adding acid.

On average, 21 pounds/9.5 kg of tomatoes, whole or halved, or 22 pounds/10 kg crushed will fill a canner load of 7 quarts/6.5 L. If you are canning 9 pints/3.5 L instead, you will need 13 pounds/6 kg, whole or halved, and 14 pounds/6.5 kg crushed.

Choose tomatoes that are disease free and vine ripened. Peel tomatoes by plunging them into hot water for 30 seconds, then plunge them into cold.

When processing large quantities, you can fill a metal colander with tomatoes and lift them in and out of a large pan of near-boiling water, then empty them in the sink that you have filled with ice cubes and water. The peel should come off in your hand. Remove the stem and discard.

You can leave tomatoes whole or halve them. Fill the jars almost full with tomatoes, then press them with a spoon to release the juice. Leave 1/2 inch/1.3 cm of headspace. Wipe the jar rim lids with a clean cloth. Put on the lids and process for 85 minutes in a boiling hot water bath, or in a pressure canner at 11 pounds/5 kg pressure, for 25 minutes. If you live at an altitude above 1,000 feet/305 m, increase water bath time one minute per each additional 1,000 feet/305 m of altitude. If you are pressure canning at a higher altitude, the time will remain the same, but you will need to increase the pressure. Check the pressure canner's manufacturer's instructions or consult with a local agricultural extension agent for the exact changes your altitude requires.

When water bath canning, bring water to a near boil before you pack the jars. You will want the water to cover the tops of the jars with 1 to 2 inches/2.5 to 5 cm of water. It's hard to gauge the right depth when you are just starting to can. Fill the pot to about two-thirds the jar height and keep a kettle of boiling water on the stove so you can add more, if needed, after the jars are placed in the water. Don't pour the water from the kettle directly on the jars, but pour it between them.

Cover the filled canning pot with the lid and begin to calculate processing time when the water reaches a rolling boil. Turn the heat down to boil gently. When the time has elapsed, use a jar lifter to remove the jars from the hot water immediately. Be sure you do not place them in a draft or extremely cold place, or the jars may crack. Allow air space between the jars so air can circulate as they cool.

For pressure canning, see the manufacturer's instructions.

When jars are cooled, test to make sure they are sealed by pressing down in the middle of the lid.

It should not move when pressed and should be slightly concave. If the lid has not sealed, you can process the jar again in the next batch, but the product will lose both freshness and flavor. The other alternative is to use those tomatoes within the next day or so, keeping them in the refrigerator until you do. The screw bands can be removed after the jars have cooled, and you can reuse them on another batch. Store the successfully canned jars in a cool, dark place.

Tomato sauces and other products may require different processing times, so process them according to the recipe.

FREEZING IS PLEASING

Frozen tomatoes have plenty of flavor, but they will not be solid when thawed so are best for soups, stews, and recipes in which you ordinarily would use crushed tomatoes.

Peel them according to the directions in the canning section and remove their stems. Quarter or chop the tomatoes and pack snugly into freezer containers, leaving 1 inch/2.5 cm of headroom. Plastic freezer bags in 1-quart/1 L sizes are great for this and can be tucked into the odd corners of your freezer. Frozen tomatoes will keep well up to six months.

A DRY BIT OF WISDOM

Garden expert Barbara Ciletti notes that a meaty paste tomato is best for drying, and she highly recommends any of these: Banana Legs, Jersey Devil, Opalka, Roma, San Marzano, or Viva Italia. Ideally they should be about 2 1/2 inches/6 cm long and 1 inch/2.5 cm thick. Barbara lives in Colorado, which has a dry climate, and she dries tomatoes in the sun using the following method:

Wash the tomatoes and remove any marks or blemishes. Do not peel. Cut the tomatoes in half lengthwise and place them on nonstick cookie sheets or a pan. Cover with a piece of cheesecloth and place in a warm outside area that has good air circulation. Morning sun is okay for sun drying, but sunny afternoons can overcook the tomatoes, and they'll stick to the pan. Turn the tomatoes every eight hours until they reach your

preferred state of dryness. If you want to store them for a long period, they must be completely dried. If you prefer a chewy consistency store in self-sealing storage bags or jars. The tomatoes must be completely dry for long term preservation. If you prefer a chewy dried tomato, be prepared to use them quickly. Or pack them in jars with olive oil, salt, and garlic, and keep them in the refrigerator.

If you live in a humid area, you may prefer to dry tomatoes in the oven. Brush them lightly with olive oil and place them on the baking sheet. Dry them at 150° F/65.5° C for at least eight hours or up to two days, turning occasionally.

You can also use a food dehydrator to dry tomatoes but they will need to be reconstituted before you use them. Dehydrators vary, so follow the manufacturer's direction.

If you have a toaster oven, you can dry tomatoes a few at a time all through the season. The toaster oven is especially good for drying cherry tomatoes and whole tomatoes that have cracked skin or another blemish. Rinse, trim, and slice the tomato(es), then arrange the slices on the toaster oven's baking sheet. If you are drying cherry tomatoes, rinse and halve. Set the gauge at 150° F/65.5° C and leave for about eight hours. You will need to keep them in the refrigerator, either in plastic bags or packed in oil as described above.

A TOMATO DUST UP

Have you noticed recently that the olive oil dip for bread in fancy restaurants comes not only garlic infused, but with a mysterious brown powder that seems to give it an almost chocolate resonance? That's very likely tomato dust, a great idea for using the tomato skins left over from peeling or canning. Place the skins on a nonstick baking sheet and pop them in the oven at 200° F/93.5° C for about two hours or until completely dry. You will want to turn them occasionally and begin checking for dryness after an hour and a half. Remove them when dry, and when they have cooled, use a mortar and pestle or spice grinder to pulverize them. Use the dust in olive oil or vinaigrette for seasoning or sprinkled over vegetables for a little extra zip.

Want to try your hand
at canning? A trio of
recipes follow to set you
on your way.

Tomato Chutney, Ketchup & Jam for Canning

Shirleigh's Red Tomato and Lemon Jam

Makes about 3½ pints/1½ L

3 cups/591 ml chopped tomatoes

2 lemons thinly sliced, with peel but with seeds removed

6 ½ cups/1538 ml sugar

1 teaspoon ground cinnamon

½ teaspoon ground cloves

6 to 8 whole cloves

8 ounces/250 ml liquid pectin

- In a large saucepan, over medium high heat, bring the tomatoes to a boil, then turn the heat lower, and simmer for 10 minutes.
- Add the lemon slices, sugar, and spices, mixing well. Over high heat, bring to a rolling boil and cook for 1 minute, stirring constantly.
- Remove from heat and stir in the liquid pectin. Skim any foam with a metal spoon. Continue to skim and stir for 5 minutes to cool the mixture and prevent the fruit from floating to the top.
- Ladle into sterilized jars and process in water bath for 5 minutes.

Special Thanks to Shirleigh Moog for her recipe.

Green Tomato Ketchup

Makes approximately 12 pints/5 1/2 L

- In a large stock pot, caramelize the onions in peanut oil over medium low heat. Add the rest of the ingredients, except the vinegar and brown sugar. Cover and let simmer for about 2 hours, stirring occasionally to prevent mixture from sticking to the pan.
- In a separate saucepan, bring the vinegar to a boil and dissolve the brown sugar in it. Add the vinegar and sugar mixture to the rest of the ingredients at the end of the cooking time.
- Use a potato masher to achieve a chunky smooth texture. Decant into sterilized jars, top with sterilized lids, and process in boiling hot water bath for 10 minutes.

5 onions, chopped
1/3 cup/79 ml peanut oil
5 jalapeno peppers, seeded and chopped
2 green bell peppers, seeded and chopped
25 green tomatoes, cored and quartered
12 garlic cloves, roughly chopped
2 tablespoons ginger powder
1 teaspoon dry mustard
1/2 teaspoon ground cinnamon
1/4 teaspoon ground cloves
3 1/2 tablespoons salt
1 teaspoon white pepper
1 cup/236 ml apple cider vinegar
1 pound/ 1/2 kg brown sugar

Yellow Tomato Chutney

Makes approximately 4 pints/2 L

- Mix all ingredients together in a large saucepan and bring to boil on medium high heat. Turn heat down to low and allow to cook until thickened to the consistency of a preserve. This will take 1 to 2 hours and you must stir the mixture frequently to keep it from sticking to the pan.
- Decant into sterilized jars, top with sterilized lids, and process in boiling hot water bath for 10 minutes. Can also be kept in the refrigerator for 2 months.

3 pounds/1 1/2 kg yellow tomatoes, cored and halved
1 cup/236 ml brown sugar
2 teaspoons minced ginger
2 teaspoons minced garlic
1/2 cup/118 ml orange juice
1/2 cup/118 ml chopped green pepper
1/2 cup/118 ml sliced red onion
3/4 cup/177 ml cider vinegar
1/2 cup/118 honey
2 1/2 teaspoons salt

Red, yellow, and green tomatoes in glass jars sparkle like jewels all winter long.

JUMPIN' JACK FROST, IT'S A GAS, GAS, GAS:

HOW YOU GET TOMATOES IN JANUARY

It may be fair to say that the tomato-eating world is divided into two parts. There are those who believe that the only tomato worth eating is one that is truly vineripe, preferably from a vine that is not so far from your table. And for anyone who has compared not only the taste but the texture, and even the color, of a tomato picked in its prime in July to those of the hard, rubbery, uniformly orange-y spheres that show up in groceries in winter, it's obvious they have a point.

Then there are those for whom the idea of a time without fresh tomatoes is unthinkable. Through the last half of the last century, it has been this latter wisdom that has dominated the tomato market. After all, what's a BL without a T? What's the point of tossing pale cukes and lettuce without chunks of red for color and heft? What's a tomato fan to do when the frost is on the vine?

WHAT'S A TOMATO FAN TO DO WHEN THE FROST IS ON THE VINE?

IT'S THE TIME OF THE SEASON

Attempts to extend the growing season to meet the commercial demand for fresh tomatoes began as early as the mid-nineteenth century. Farmers' almanacs and horticultural journals began focusing on how to get an early start and on breeding tomatoes to favor early producers. The reason why was clear: The first farmer to get to market with a ripe tomato could command a premium price. The later in the season, the greater the bounty of fresh tomatoes available and the cheaper they would sell per unit.

It was also at this time that southern tomato producers discovered that if they shipped their product to the north before the season there began, the price would go up enough to offset the cost of transportation. So farmers began to breed tomatoes with an eye to travel-worthiness; in other words, with firmer flesh and thicker skin. Even so, there were still plenty of times of the year when you simply couldn't get a fresh tomato in the market.

With the creation of the first commercial hybrid in the mid-1940s and the rapid improvement in transportation methods in the latter part of the twentieth century, however, it eventually became possible to get a tomato from any "here" to just about anywhere else. Getting it there economically and in one piece, however, required some tinkering.

SIX DAYS ON THE ROAD

First there was the matter of breeding tomatoes for travel, not taste. The very quality that makes a tomato picked from the vine such a pleasure to eat–its bursting ripe juiciness–is an insurmountable obstacle in shipping. The first solution was to breed tomatoes that were not only firmer of skin but firmer of flesh and less juicy inside.

Commercial breeding also focused on creating tomatoes of a more uniform, rounded shape for ease in mechanical harvesting. (Notice that many of the handpicked heirlooms you'll find in the farm market have crenellated crowns akin to the early tomato's shape.) Tomatoes were bred with a focus on selecting those that were most disease-resistant and those that could hang around on a shelf for the longest period of time. The one thing that wasn't given much consideration in this selective breeding was taste.

Eventually tomato producers realized that the greener the tomato was when picked and shipped, the more stable it was for travel. Aware that the tomatoes naturally produced ethylene gas as their ripening agent, the fresh tomato industry began harvesting green tomatoes for shipping, blasting them with ethylene later in the process to induce reddening.

You may assume that when you see a tomato in the market labeled "vine ripened" it means that tomato hasn't undergone this process, and was, in fact, picked when it was fully ready to eat. It

ain't necessarily so. First, there are no legal regulations governing the label "vine ripened." Within the industry, however, it's understood that this should mean a tomato that has at least started to show a pink color before it's plucked from the vine, and has not had gas used to induce further ripening. In reality, not all of the tomatoes under the vine-ripe label meet these criteria; and even those that do won't have the same intensity of flavor as a fruit that is allowed to reach its maturity, or near maturity, on the mother plant.

Why is that? If you've watched your green tomatoes growing on the vine, waiting for that perfect moment to pick a juicy, tasty one for frying, you will know that the color of green begins to change in the last days before ripening, becoming increasingly lighter. That happens because the chlorophyll inside has started to break down, the first step toward ripening. This is the stage when most commercial tomatoes get plucked and packed for the market. Technically, once this process has begun, it doesn't stop and the tomato will continue to turn red and, because its sugar content is increasing and its acid content is decreasing, it also will develop a taste more akin to that of a ripe tomato than a green one. In order to develop fully, however, the tomato must also have sunshine and heat during this period or the whole complex spectrum that is a ripe tomato's distinctive taste simply won't happen. Sunshine and heat are two things a commercial green tomato waiting to go to market isn't going to get.

Taste isn't the only thing lacking in these prematurely plucked tomatoes. Nearly half of the vitamin C and beta-carotene that is present in a genuinely vine-ripened tomato is missing in one of equivalent variety and size that has been picked at this mature green stage and ripened with ethylene.

LOVE MAKES STRANGE BREAD FELLOWS

So, okay, no one needs to tell you that your Granny's garden tomato was not only sensually but morally superior to those gassed up tomato wannabes at the supermarket. But even so (whine, whine), there are just some times when a guy or a gal has to have a burger with all the trimmings, even in the dead of winter.

So here's what you do: Use cherry tomatoes–year round they have an acceptable level of tomato bite. And cherry tomatoes travel well, so chances are you'll be getting some that were grown somewhere that tomatoes really are in season. Likewise, plum tomatoes will have a better flavor than the standard rounded ones, even when they are not drop-dead ripe.

And more good news: Hydroponic tomatoes are now available year round in most markets, and exhibit markedly better color and somewhat better taste than other hothouse varieties. (See page 112 for more information on hydroponics.) But they usually come with a pretty high sticker price.

If these "fresh" winter options still won't satisfy your hunger, you can try to get your tomato hit from a can. Think "marinara," "salsa" or pasta with chunks of canned tomatoes cooked with plenty of garlic. No, most canned tomatoes won't transfer to a sandwich or salad–the texture is all wrong and the flavor is often too salty and, unless cooked, a little tinny. I say "most," however, because lately cans of fire-roasted tomatoes have been appearing in the organic food aisles of my local market, and they actually can provide an interesting alternative to fresh in a couple of cases. Buy the diced fire-roasted tomatoes and drain them, reserving the juice for your next pot of soup (or drink it as a beverage). You can sprinkle the chunks of tomato over bacon on toast and top with

lettuce, on a burger, in a pita with felafel, or on a ham sandwich. The texture won't work with a tossed salad, but they are a fine substitute in a pasta salad that usually calls for fresh tomatoes or in guacamole. Nope, the taste is not like that of a garden tomato, but it still has plenty of tomato-y zing and zip, and the fire-roasting adds a smoky undercurrent that creates some complexity in the flavor and masks any underlying tin can taste.

WATER MATERS: HYDROPONICS 101

Hydroponic tomatoes are appearing with increasing regularity in groceries around the world, but while many consumers have sampled them, not everyone knows what they are.

Hydroponics is a method of growing plants in nutrient solutions (water and fertilizer) with or without the benefit of a support medium such as sand, gravel, vermiculite, peat, or sawdust, Virtually all of the commercial hydroponic production of tomatoes in the world is being conducted in enclosed greenhouse space.

Tomato plants cultivated in this fashion exhibit dramatically increased growth; they reach up to 40 feet (about 12 meters) long and bear fruit every few inches. Plants are generally strung up on wires to run through the greenhouse. Some producers extend the growing season by planting a second succession as the first season plants begin to decline. Because the plants don't require soil, pots, etc., and because the growth and yield are so increased, it's possible to produce considerably more tomatoes at one time than with regular hothouse methods.

One of the obvious advantages of hydroponic production is that it doesn't depend on optimum climate or outdoor growing conditions, therefore hydroponic growers can provide tomatoes to the market in the winter months and at premium prices.

Too much of a premium for many consumers, however. One of the drawbacks of hydroponic growing is the high cost of the start-up technology and of the labor-intensive care needed to maintain the production. In addition, hydroponic tomatoes are often shipped great distances, frequently by air, to reach the prime markets. This, too, adds to the price, making hydroponic tomatoes sometimes as much as three times the price of an ordinary tomato.

The tomatoes generally have a redder color than other out-of-season tomatoes, and a better fragrance. But they fall well below summer tomatoes in the taste department. Consequently consumers have been reluctant to purchase them with regularity, and the market hasn't yet stabilized. Hydroponic growing on a large scale is still in the early stages, however, and research into the impact of nutrients on taste and the creation of specific hydroponic varieties with an emphasis on flavor as well as travel-worthiness could yield interesting results in the future.

Firm winter tomatoes work well stuffed and cooked in these next recipes.

A typical hydroponic tomato greenhouse.

Hydroponic tomatoes generally have a redder color than other out-of-season tomatoes, and a better fragrance.

Spinach & Artichoke Stuffed Tomatoes

Serves 8

1/3 pound/150 g fresh spinach
1 cup/236 ml sour cream
2/3 cup/158 ml mayonnaise
1 cup/236 ml canned artichokes, drained and chopped
1/2 teaspoon garlic salt
1/2 pound/226 g Swiss cheese, grated
8 large firm tomatoes
salt
1/4 cup/59 ml bread crumbs

- Steam spinach until tender, about 3 minutes. Remove and press to drain water. Chop into small pieces.
- Mix together spinach, sour cream, mayonnaise, artichokes, garlic salt, and Swiss cheese until thoroughly blended. Cover and refrigerate over night.
- Prepare tomatoes for stuffing as described on page 6.
- While tomatoes are draining, preheat oven to 425°F/218°C and remove spinach artichoke filling from refrigerator to bring to room temperature.
- Lightly oil a baking sheet large enough to accommodate all the tomatoes without crowding. If you don't have a single sheet large enough, use two.
- When tomatoes are drained, arrange on the baking sheet(s). Use a spoon to fill the tomatoes to the top with spinach artichoke filling. Sprinkle with bread crumbs.
- Bake in oven for about 20 minutes, until filling is bubbly and bread crumbs are lightly browned. Remove and serve warm.

Zesty artichoke and spinach gratin is just the stuff for turning plump tomatoes into brunch!

Orzo, Olive & Feta Stuffed Tomatoes

Serves 5

5 large, ripe tomatoes
3 cups/354 ml cooked orzo
1/4 cup/59 ml sliced black olives
1/4 cup/59 ml finely diced red onion
1/4 cup/59 ml diced roasted red peppers
2 teaspoons minced fresh mint
1/2 cup/118 ml goat's milk feta cheese
1 tablespoon olive oil
salt
pepper

- Preheat oven to 350°F/176°C. Lightly oil a baking sheet.
- Prepare tomatoes for stuffing as described on page 6.
- Thoroughly combine all the other ingredients in a mixing bowl, adding salt and pepper to taste.
- When tomatoes are ready, stuff them with the orzo mixture and bake at 350°F/176°C for 20 minutes. Serve warm.

Fragrant and savory right out of the oven, these yummy tomatoes will warm your heart.

Early Girl Marinara Sauce

Makes 1 gallon

- In a large, heavy stockpot on low heat, cook the onions in the oil until they just begin to soften. Add the carrots and garlic and simmer for 5-10 minutes to let the flavors blend. Add the fresh tomatoes and fresh herbs and simmer for 20 minutes. Add all the rest and simmer, uncovered, for 5 hours, stirring frequently to keep mixture from sticking.
- Marinara can be frozen in smaller containers up to 6 months after preparation. It can also be canned.

5 large onions, chopped
2 cups/473 ml olive oil
3 cooked carrots, pureed
1 cup/236 ml pureed garlic
10 paste tomatoes, pureed
5 tomatoes, diced
1/2 cup/118 ml chopped fresh basil
1/2 cup/118 ml chopped fresh oregano
96 ounces/3 L canned tomatoes
6 ounce/170 g can tomato paste
5 bay leaves
12 cups/1.5 L water
1 1/2 cups/354 ml red wine
1/3 cup/79 ml brown sugar

Mama mia! Such a wonderful alchemy when tomatoes simmer for hours!

HERE'S TO HEALTH:

THE TOMATO IN MEDICINE

Lift your glass to lycopene, the pigment that gives most tomatoes their red color and an antioxidant with a proven track record of health benefits. But make sure it's a glass of heat-processed tomato juice you hoist, to insure you get the most of lycopene's merit.

At the turn of this century researchers announced the results of several studies that show that the consumption of tomato products–particularly cooked ones–could have a preventative effect on a number of cancers and other diseases. In the late 1990s, Harvard University released the results of a six-year study that showed that men who ate at least 10 servings of food containing tomato sauce or tomatoes per week were 45 percent less likely to develop prostate cancer. The positive agent was determined to be lycopene, which acts as an antioxidant in the body.

Since then studies into lung, bladder, cervical, skin, breast, and ovarian cancers have yielded promising, though not yet conclusive, results. In addition, it was discovered that lycopene has the ability to lower bad cholesterol levels and is associated with a lower risk of heart disease. There is some indication that it can also help prevent arteriosclerosis and cataracts. If the latter proves true, it would confirm the recommendation of early herbalists who, as far back as the sixteenth century, prescribed the juice of the tomato as a cure for infections of the eye and cataracts.

But raw juice may not be your best bet for getting the most from tomato's curative punch. In fact, this is one case where, nutrition wise, fresh is not better than cooked. Studies have shown that while there is plenty of lycopene present in a raw tomato, it's not readily taken in by the body. Preparing the tomato, by mashing and cooking, appears to break down cells in such a way that the body can more easily absorb the lycopene, up to four times as much compared to raw. There is also some indication that a moderate amount of fat consumed with the tomato may help in lycopene absorption, as well. (Other foods that are lycopene rich include watermelon, red grapefruit, and carrots.)

Of course, you don't want to eat all of your tomatoes cooked. One hundred milligrams of raw tomato provides about 40 percent of the RDA for Vitamin C, which would be destroyed if the tomato were heated. It's also a great source for Vitamin A (30 percent of the RDA), either fresh or cooked. And the tomato contains small amounts of iron, potassium, calcium, sodium, thiamin, and riboflavin.

THE WAR OF THE TOMATO MEDICINE MEN

This explains why even before these most recent discoveries, the tomato has had a reputation for being a health food. In fact, the purported curative qualities of the fruit led to what historian Andrew F. Smith has dubbed The Great Tomato Mania in the United States during the 1830s and '40s. During this time, doctors touted the tomato as a cure for diseases such as dyspepsia, scrofula, and consumption. Miracle cure stories abounded including some in which cholera was remedied by the ingestion of quantities of fresh tomatoes.

In the midst of the mania, a spate of patent medicines surfaced purported to contain extract of tomatoes; and a veritable war of words that rivaled the worst political mudslinging developed between two of the most prominent medicine manufacturers: Archibald Miles and brothers Guy and George Phelps. Smith's excellently researched book, *The Tomato in America* contains a detailed and fascinating account of the quarrel and the mania that inspired it. Smith suggests that some of the early claims for the tomato's healing properties were

not unfounded. Noting that the American diet of the era was heavy on meat and fatty and fried food and noticeably lacking in vegetables, Smith suggests that the vitamins and minerals available in tomatoes may well have improved the health of many who succumbed to the mania.

Here are some healthy recipes to make the most of the tomato's curative powers.

Fresh Tomato Juice

Serves 2

10 medium-size tomatoes
1 teaspoon sugar
1 teaspoon salt
hot pepper sauce
black pepper

- Core and quarter the tomatoes.
- In a heavy saucepan over medium low heat, simmer the tomatoes, sugar and salt for 30 minutes. stirring and mashing frequently. Strain through a food mill. Discard the pulp and chill the juice.
- When you are ready to serve, add hot pepper sauce and ground black pepper to taste.

Tomato, Tahini, Tamari Dressing

Makes 1 cup

1/2 cup/118 ml very juicy tomatoes
1/6 cup/36 ml tahini
1/8 cup/29 ml tamari soy sauce
1/8 teaspoon dried oregano
1/8 teaspoon garlic powder
1/8 teaspoon dried parsley
1/8 teaspoon dried basil
pinch of salt

- In a blender, process the tomatoes at medium speed for 1 minute. Add the tahini, tamari, and seasonings and blend again until pouring consistency. If the mixture is too thick, you can add a little tomato juice or water. Refrigerate until ready to use on your favorite vegetable or noodle salad.

Fattoush

Serves 4

- 1 tablespoon olive oil
- 1 tablespoon champagne vinegar
- 2 tablespoons chopped fresh cilantro
- 2 tablespoons chopped fresh parsley
- 2 tablespoons chopped mint
- 1 teaspoon ground sumac
- 2 cups/473 ml seeded and diced tomatoes
- juice of 2 lemons
- 1/2 cup/118 ml diced green pepper
- 1 cup/236 ml peeled, seeded, and chopped cucumber
- salt
- 1 pita bread, crisply toasted

❖ In a bowl, combine all ingredients except the pita bread, salting to taste. Chill for at least 1 hour.

❖ When you are ready to serve, break the pita bread into thumb-size pieces and toss with the rest of the ingredients. Serve immediately.

Tomato & Anise Pork Roast

Serves 6

- 1 cup/273 ml chicken stock or water
- 4 cups/946 ml diced onion
- 6 to 8 pound/2 1/2 to 3 1/2 kg pork roast
- 2 cups/473 ml tomato puree
- 2 star anise, ground
- 1/2 cup/118 ml brown sugar
- 2 tablespoons minced fresh ginger
- 2 cloves garlic, minced
- 1 tablespoon soy sauce
- 1/2 tablespoon rice wine vinegar
- 1/2 cup/118 ml red wine
- 1 teaspoon crushed red chili

❖ Preheat oven to 300°F/149°C.

❖ Pour the stock into the bottom of a large, ovenproof pot with a lid. Arrange the onions evenly on the bottom and place the pork roast on top.

❖ In a mixing bowl, combine all the remaining ingredients and stir together to make a thin sauce. Pour the sauce over the roast and cover.

❖ Cook at 300°F/149°C. for 2 1/2 hours. Remove from pot and allow to rest for 10 minutes before slicing. Serve covered with the sauce from the pot.

CANNERY ROW:

PROCESSED TOMATOES

The tomato can. Think of it as a verb, like this:

The tomato can travel anywhere regardless of the roughness of terrain or the cramped conditions of transport.

The tomato can be eaten any time of year.

The tomato can be sitting close at hand in your kitchen to become a meal-in-minutes as soup, sauce, or when added to other ingredients.

The tomato can be processed in such a way that it is still one of the most economical buys in the grocery.

Now you begin to see why the tomato is the most popular processed fruit/vegetable in the world.

For the last several years, production has held steady at close to 22 million metric tons in the 11 major producing countries. The United States leads the world in both production of processed tomatoes and consumption, and Italy is second.

But it is a Frenchman who is credited with processing the first tomatoes commercially.

Nicholas Appert began bottling experiments in France at the turn of the nineteenth century. In one, he extracted skin, seeds, and liquid, leaving tomato pulp, which was then put up in bottles and sealed in a hot water bath. His processes were soon translated and available in both

Britain and the United States. It wasn't long until tinned tomatoes and bottled tomato ketchup followed.

The use of metal cans for preserving tomatoes began with the Royal Laboratory in Greenwich, where food was put up in tins as early as the middle of the eighteenth century. The first commercial food canning operation began in Britain under the aegis of Brian Donkin and John Hall, businessmen who were canning meat, vegetables, and soup for sale by 1814. Canned food became essential for the British Navy and was sold elsewhere as a gourmet item. Commercial canning came to the United States with British immigrants of this era. But while many foods were put up in metal containers, tomatoes were still generally canned in glass.

This changed in large part through the efforts of Harrison W. Crosby, who not only processed tomatoes in tin pails but worked assiduously to create a market for the product. He sent samples to Queen Victoria, President James K. Polk, U.S. senators and newspaper editors, and restaurants and hotels. The tomatoes caught on just as the California Gold Rush began, and they became a high-ticket necessity shipped to the American West.

WAR? WHAT IS IT GOOD FOR?

Had Edwin Starr been around in 1865 to ask that rhetorical musical question, he might have been surprised at the answer. The U. S. Civil War was, among other things, significant in boosting the American canning industry. The Union Army made ample use of tinned food for rations, and the Confederate forces made good use of Union stores when they won a victory. By the end of the war, both sides had acquired a taste for canned food, particularly tomatoes.

During the war, growers in the Northeast had stepped up tomato production since shipments of fresh tomatoes from the South had stopped. With the canneries already in place, the tomato became the prime candidate for processing in the postwar period. The canned tomato became not only a staple in the kitchens of the era but also a necessity for the settlers now moving west. Canned tomatoes were sold in the mining camps of the Rocky Mountains and cans littered the trails of the wagon trains across the plains and through the mountains. A traveler writing in 1865 noted canned tomatoes "in every hotel and station meal and at every private dinner or supper."

HOW DO WE LOVE THEE, CANNED TOMATO?

Let us count the ways. The tomato took supremacy as a canned item in large part because of its versatility as a product. Compare the tomato to canned pineapple, for instance, which you can find in slices or chunks, crushed or juiced. Tomatoes come in all of those configurations and whole as well. They are also used to make a vast variety of sauces for ethnic foods, canned with chilies or other vegetables added, and serve as the main ingredient in a number of canned soups. Tomatoes form the base of most salsa and ketchup, the number one and two most popular condiments in the world. This versatility not only means the tomato can be put up in a number of marketable configurations, it also means that precious little is lost in the processing of the food.

Canned tomatoes were also the product of that old economic saw: supply and demand. Eager to get the first tomatoes to market to command the premium early season price, farmers filled their fields with tomato plants to get as many early fruits as possible. By the time the midseason hit, however, everyone had so many tomatoes that the market price had fallen. Canneries stepped in to buy the fresh tomatoes at low, low prices, assured that the canned product would sell for much, much more when the fresh season had ended. Commercial tomato canning was also a relatively low cost process.

The first canneries in the United States were located in the northeastern part of the country, which dominated the market until the 1930s or so when Midwestern canneries began to spring up. Purdue University tomato expert Phil Nelson notes that there were some 200 tomato canneries in the Midwest in the 1960s, but soon they began moving to be near the massive agricultural fields of California, Now there are only five canneries in the Midwest. The international tomato-processing field currently is dominated by California.

TOMATO LABEL ART

American painter Robert Henri, one of the most influential voices in the American Independent Art Movement, wrote *The Art of Spirit* in 1923. In this collection of notes, theory, instruction, and memoir, he recalled advising a young artist concerned about making money from his particular style of work. Henri wrote:

"He happened to be a man of considerable talent and had great enthusiasm in his work. But I knew there was no public enthusiasm for such work. I remembered he had told me that before he got really into art he had made a living by designing labels for cans, tomato cans and the like. I advised him to make tomato-can labels and live well that he might be free to paint as he liked."

It was 35 years later that Andy Warhol turned Henri's advice on its ear and created the famous Campbell's Tomato Soup can image that came to epitomize Pop Art. Posters of Warhol's work sell on the Internet now for less than $75 a pop. In other spots along the e-commerce highway, however, collectors are buying actual labels that once were printed to surround real tomato cans and treating them as art.

These labels are more than mere kitsch, valued for their novelty. Some of them are lovely exam-

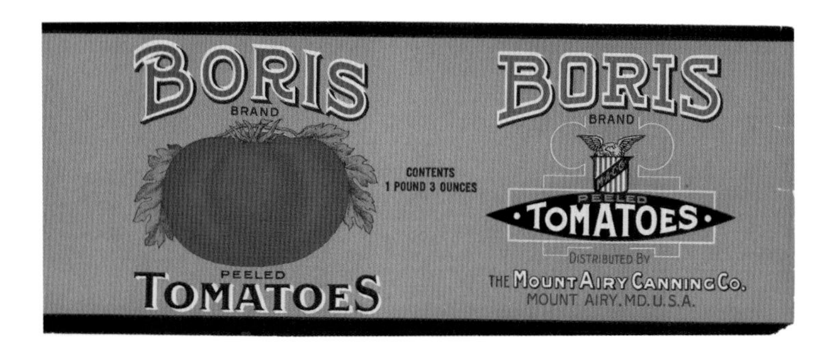

ples of proletarian art. Practical and informative, they nevertheless manage–with their plump, red spheres, curling vines and decoration–to create a romantic picture of the contents inside, one geared to the particular time and place from which they arise.

Who could miss that Little Miss, with the neatly coifed woman in girlish necktie, was aimed at the "new" housewife, concerned with nutrition and efficiency? Or that Red Barn Pole Tomatoes promised to deliver the pleasures of rural living to your kitchen in a tin?

Other labels are simply beautiful on their own, miniature works of art that may make you wonder if Henri's young protégé might have been the eye and hand behind their crafting.

RED BARN POLE TOMATOES PROMISED TO DELIVER THE PLEASURES OF RURAL LIVING TO YOUR KITCHEN IN A TIN.

Masala Base

Makes 1 quart/1L

1 cup/236 ml vegetable oil
5 large onions, chopped
1 tablespoon minced fresh ginger
½ tablespoons ground cumin
½ tablespoons ground coriander
¼ tablespoons ground red chile
½ tablespoon salt, plus
1 head of garlic, peeled and chopped
6 cups/1416 ml chopped tomatoes
6 cups/1416 ml water

- In a heavy pot over medium low heat, brown the onions in the vegetable oil. Add the fresh ginger, ground spices, and salt and sauté 2 to 3 minutes.
- Add the garlic, tomatoes, and water and simmer for 1 hour on low heat, stirring occasionally. Take care not to let the mixture burn. Taste and add more salt, if needed.
- Base can be frozen in smaller containers and used for Channa Shag (page 130), Curried Potatoes (below) or other Indian recipes.

Curried Potatoes

Serves 4 to 6

2 cups/437 ml Masala Base(above)
3 cups/708 ml diced boiled potatoes
1 cup/236 ml frozen peas
½ tablespoon ground ginger
1 tablespoon ground curry powder
1 tablespoon garam masala
1 tablespoon honey

- In a heavy saucepan over low heat, simmer the Masala Base until it's steamy but not boiling. Add the potatoes, peas, ginger, curry, garam masala, and honey. Simmer for 20 minutes and serve warm.

Tomatoes and spice make these Indian dishes so nice.

MASALA

Masala is the term used for different blends of Indian spices. Garam masala is one of the most popular and can be found prepared in Indian markets, gourmet stores, and some groceries. The Masala Base recipe is a spicy foundation for any number of dishes, including the two here, Channa Shag and Curried Potatoes. The fundamental taste is subtly changed by the addition of other, different spices depending on the dish.

Channa Shag

Serves 4 to 6

2 cups/473 ml Masala Base (page 128)

1 cup/236 ml cooked, chopped spinach

1 cup/236 ml drained and rinsed cooked chickpeas

1/2 tablespoon ground curry powder

1 tablespoon ground garam masala

4 cups/946 ml cooked basmati rice

- In a heavy saucepan over low heat, simmer the Masala Base until it's steamy but not boiling. Add the spinach, chickpeas, curry, and garam masala.
- Simmer for 15 minutes to allow the flavors to mingle. Serve over rice.

Shrimp Creole

Serves 8 to 10

2 tablespoons olive oil

2 cups/473 ml onion, diced

1/2 cup/118 ml red pepper, diced

1/2 cup/118 ml green pepper, diced

1 cup/236 ml celery, diced

1/8 cup/29 ml red wine

1 cup/236 ml vegetable or chicken stock

2 1/2 cups/591 ml tomato puree

1 tablespoon brown roux (SEE NOTE)

salt

pepper

1/2 pound/1/4 kg fresh shrimp, peeled and deveined

4 cups cooked rice

- In a heavy pan with a lid, on medium low heat, sweat the onions, peppers, and celery in olive oil until the onions turn translucent. (To sweat, put all the ingredients into the pan at the same time, stir and cover, so all the steam stays in the pot.)
- Remove the lid, add the red wine, and turn up the heat a bit. Cook at a lively simmer for a few minutes to reduce the wine.
- Add the stock and tomato puree and bring to boil. Whisk in the roux and let simmer for 30 minutes, stirring occasionally to keep from sticking. Add salt and pepper to taste.
- Toss in the shrimp and cook for a few minutes until shrimp are done. Serve over rice.

NOTE: Prepared *roux* is available in most gourmet stores and many groceries. If you want to make your own, melt 1 cup/236 ml of butter in a heavy saucepan. Sprinkle 1 cup/236 ml white flour over it and stir with a wooden spoon constantly until the mixture turns a deep, golden brown.

Roux is the characteristic thickener used in Creole and Cajun dishes, and is useful in other recipes. You may refrigerate in a tightly sealed jar for several weeks.

COSTARRING ROLES:

KETCHUP AND SALSA

Diced, whole, crushed, and made into paste, canned tomatoes are ubiquitous in the modern kitchen. But tomatoes aren't satisfied to appear unadorned on the pantry shelf. Instead they have finagled their way into two of the most beloved food items in the world: tomato ketchup and salsa.

Ketchup didn't start out as a tomato product. In fact, it didn't start out as ketchup. The precise etymological origins of the word are sketchy, at best, with various theories suggesting that it evolved from words of English, Spanish, Portuguese, Arabic, Japanese, or Malaysian origin. The Oxford English Dictionary says that it derives from ke-tsiap, a word with Chinese roots meaning "the brine of pickled fish," and this has become the most accepted, though hardly universal, theory.

Although ketchup has been around for centuries and is used around the world, it still has more than one English spelling. In addition to ketchup, it is variously spelled catsup or catchup, although the "k" spelling is becoming preferred. An 1831 tract called the Domestic Chemist opined that the words were used to "indicate a sauce of which the name can be pronounced by everybody, but spelled by nobody."

At that time it might also have been said that the sauce could be eaten by anyone, but its ingredients specified by no one. That's because early ketchups were made from a variety of items, including anchovies, oysters, mushrooms and walnuts. There was even one made of kidney beans. Tomato ketchup was a late entry into the field.

THE TOMATO PLAYS CATCH UP

A recipe for English "Katchop" appearing in 1727 was the first published and featured anchovy, vinegar, wine, and many spices but no tomatoes. Variations of this recipe, plus those for mushroom and walnut ketchups, dominated eighteenth century cooking in England, and traveled from England to America. The first written tomato ketchup recipes appear in the United States between 1795 and 1800, so it's fair to assume that tomatoes were being used to make the sauce at least a decade or two prior to their publication. Although the terms tomato sauce and tomato ketchup were sometimes used interchangeably early on, the sauce was usually intended for quick use. Ketchup, on the other hand, was meant to be stored and so evolved to contain a quantity of vinegar, sugar, and more spices than an ordinary sauce.

Ketchup was used as a condiment, much as we use it today, but it was also considered an essential ingredient in the preparation of more complex dishes. Although it has become fashionable to look down our noses at the idea of spiking a soup, sauce, or pot of chili with something so mundane as ketchup, the truth is that this common prepared product contributes a surprising complexity to many a fancy dish.

BUT DOES A GREAT KETCHUP HAVE LEGS?

In recent years several publications have set up ketchup tastings to determine the best of the commercial brands. One sampled ketchup in spoonfuls, another on crackers with soda water to clear the palette. *Vogue* food writer Jeffrey Steingarten put the tomato product to the real test, however, tasting 30-some ketchups over several days slathered on McDonald's french fries and washed down with Coca-Cola. The tastings have had various results, but over all, a definition of what constitutes a great ketchup has emerged. A strong red color is the only way to go. It must have a good balance of the sweet and piquant and a mild but not overpowering saltiness. Likewise, the spices should contribute to its complexity and flavor without being so strong that they are individually identifiable. The texture can be smooth or slightly lumpy, depending on the diner's preference, but the ketchup should never separate on the plate leaving a pink, watery runoff. The aftertaste should linger and leave a pleasant sense of fullness on the tongue, but it should not be cloying.

Ketchup is an international condiment, and depending on where you are, it may be sweeter, hotter, spicier, or appear with curry or paprika as a primary ingredient. In Japan it is eaten on cabbage roles, as well as on hot dogs and fries, and is used to top pasta, as it is in Holland, Venezuela, and Greece.

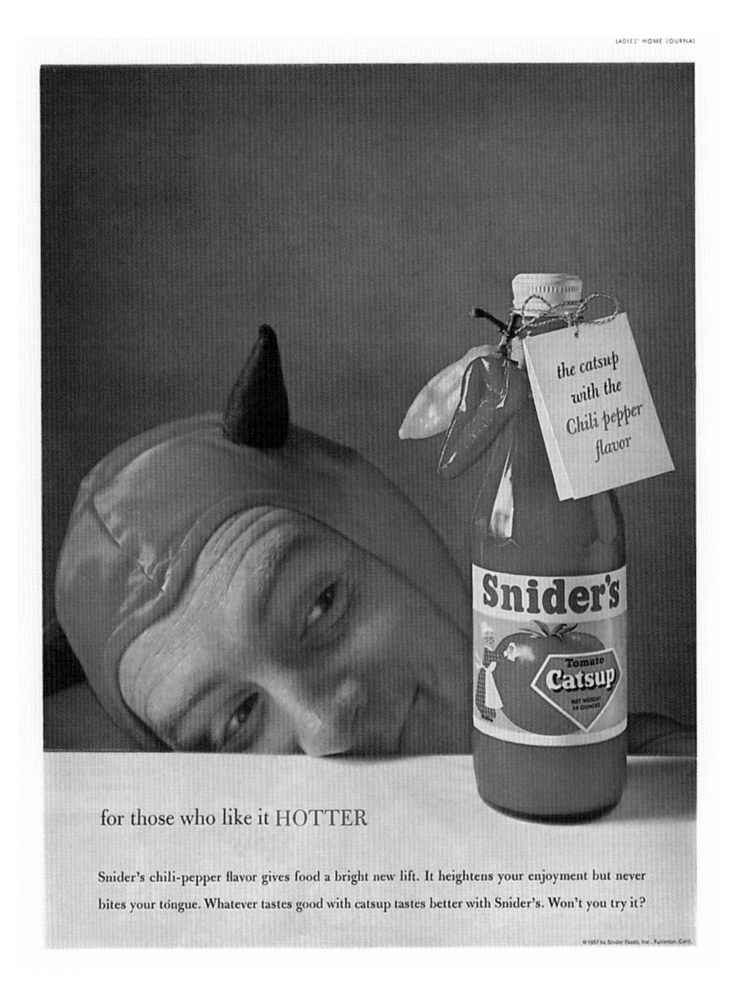

SASSY SALSA RACES DOWN THE STRETCH

Dollar for dollar, salsa has outstripped ketchup in international sales since the early 1990s, but ounce for ounce, ketchup is still condiment king. That's because the price for commercial salsas is higher per unit than for ketchup, but ketchup is still consumed in greater quantity. Salsa's gains in the market share, however, have been significant and continue to grow. Market analysts may be inclined to call it the latest thing, but in fact, salsa is the oldest known tomato recipe of which there are reports. A Franciscan priest in Mexico in 1529 noted that the natives there combined

fresh tomatoes with chilies and ground squash seeds to produce a sauce that was quite tasty on fish, lobster, sardines, turkey, and venison. This versatility is one reason salsa has become a mainstay in the pantry. But equally important is its role (paired with corn chips) as an instant snack or appetizer for unexpected guests.

IS THE BEST SALSA THE SIMPLEST?

In a rush to get a piece of the burgeoning market, hundreds of boutique salsa makers have sprung up, many capitalizing on local ingredients, others aiming for the dramatically unusual. Salsa does not have to have tomato as its primary base, and mango, papaya and watermelon varieties have started to show up with regularity in top restaurants. On the grocery shelf, it is possible to find salsa with every imaginable kind of pepper featured, including plain green pepper for the timid of taste. Exotic combinations also show up, sometimes to no good end. I once bought a jar of salsa made with green chile, tomatillos, and white chocolate thinking that such an odd combination had to be good or why do it? I was wrong, wrong, wrong. Sometimes there are reasons a particular recipe has remained relatively unchanged for, oh, nearly five centuries. It's not without reason that the absolute best salsa you can make is also one of the simplest, depending only on the season to be perfect. That season should be the very end of summer when tomatoes hang from the vine in weighty succulence. Pick the juiciest one you can

find, procure a sweet white onion (the flatter, Vidalia type), and whatever fresh green chile is best in the market. You're golden if you happen to live near the Hatch Mesilla River Valley in southern New Mexico where the greatest green chilies on earth are grown.

Roasting the pepper makes its flavor much richer and more complex. I roast mine in the toaster oven, on the tray with the temperature set to toast. I turn them after each cycle so that the skins char on every side. You can also roast chiles on an outdoor grill (while you're cooking something else, as well, if you like) or over the open flame of a gas stove, holding the pepper speared on a fork.

When the skin is charred, place the pepper in a brown paper bag and let it sit for 15 minutes. Remove it from the bag and rub lightly between the palms. The skin should come right off. Remove the stem and seeds, and chop the chile into pieces about the size of the nail of your little finger. Taste it and determine if you will need more to make your salsa as hot as you want. Set the desired amount of chile aside in a medium-size bowl.

Peel the tomato and chop it in pieces a little larger than the chile. Add to the bowl with all of the tomato juice. Chop the onion in pieces the same size as the chile, enough to make about one half cup. Stir it all together and salt to taste. Nothing more is needed except, of course, a bag of chips.

If you'd rather make your own...

Homemade Ketchup

Makes approximately 4 pints/½ L

10 pounds/5 kg fresh tomatoes, cored and chopped
2 tablespoons olive oil
3 onions, chopped
2 cups/483 ml red bell peppers, chopped
1 teaspoon ground cinnamon
1 teaspoon paprika
2 teaspoons celery seeds
½ teaspoon ground allspice
½ teaspoon ground cayenne pepper
1 cup/236 ml apple cider vinegar
½ cup/118 ml brown sugar
½ teaspoon salt
½ teaspoon ground black pepper

- In a heavy pot on medium heat, cook the tomatoes in the olive oil until thoroughly stewed, about 40 minutes.
- Run the tomatoes through a food mill and discard the seeds and skins. In a blender or food processor, puree the onions and peppers with 2 cups/473 ml of the milled tomato mixture until smooth.
- Return the milled tomatoes and the vegetable puree to the pot and add the cinnamon, paprika, celery seed, allspice, and cayenne pepper, and stir to combine well. Simmer for 1 hour.
- Add the apple cider vinegar, brown sugar, salt, and pepper, and stir to combine well.
- Decant into sterilized jars, top with sterilized lids and process in boiling hot water bath for 20 minutes.

KILLER TOMATOES

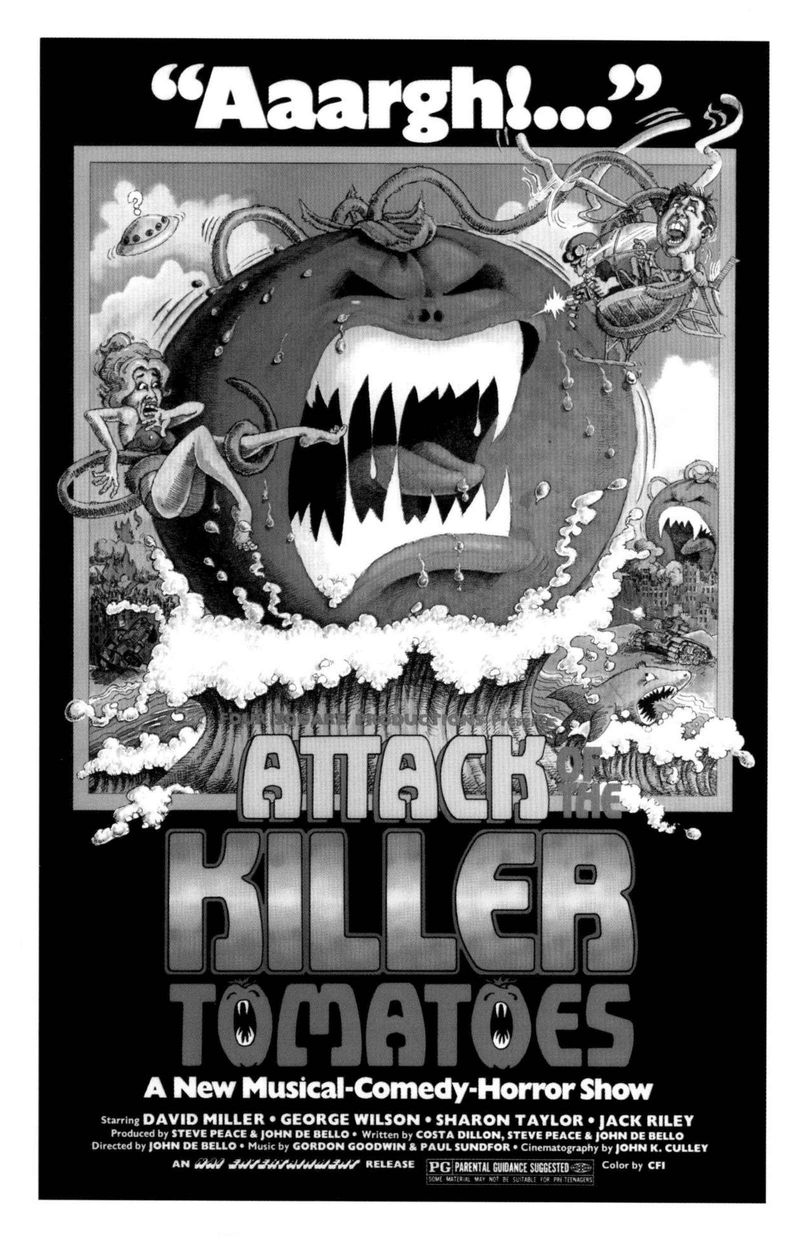

"Aaargh!..."
ATTACK OF THE
KILLER
TOMATOES
A New Musical-Comedy-Horror Show
Starring DAVID MILLER • GEORGE WILSON • SHARON TAYLOR • JACK RILEY
Produced by STEVE PEACE & JOHN DE BELLO • Written by COSTA DILLON, STEVE PEACE & JOHN DE BELLO
Directed by JOHN DE BELLO • Music by GORDON GOODWIN & PAUL SUNDFOR • Cinematography by JOHN K. CULLEY
RELEASE
PG PARENTAL GUIDANCE SUGGESTED
SOME MATERIAL MAY NOT BE SUITABLE FOR PRE-TEENAGERS
Color by CFI

Like a genetically modified tomato experiment gone awry, director John De Bello's 1978 film, *Attack of the Killer Tomatoes!* (possibly the worst intentionally bad movie ever made) has spawned offspring: *Return of the Killer Tomatoes!* (1988); *Killer Tomatoes Strike Back!* (1991); and *Killer Tomatoes Eat France!* (1992). There is even a director's cut available.

What is it about this film, universally deemed as an attempt at parody so dreadful it's not even funny for its failure, that keeps both audiences and director coming back? Not surprisingly, the place to define the movie's mystique is the on-line movie review and rating web site called, appropriately, Rotten Tomatoes.

DID YOU HEAR A SPLAT?

Created by movie buff Senh Duong in 1998, the web site has more than 100,000 titles and 360,000 review links in its database and is growing like a hydroponic tomato vine. Its name echoes the tradition of throwing rotten tomatoes during live performances of particularly odious nature. Certain approved critics on the web site's list are asked to rate the movie in question as either fresh (a nice red tomato icon) or rotten (an ugly green splat). These ratings are averaged together to come up with a percentage on the Tomatometer indicating where the film stands.

The original *Attack of the Killer Tomatoes!* rates 29 percent; in other words, five out of seven Approved Tomatometer Critics give it a splat.

The two critics who didn't declined to post reviews saying why they thought it deserved a better rating. One could speculate their fondness may have been sparked by stellar dialogue such as this:

"No weapon, no motive, no clues. All we have to go on is this bloody corpse."

"Look again, Harry. That's not blood, its . . . tomato juice!"

ROOTS AND BRANCHES

The movie begins with a nod to Alfred Hitchcock's classic, *The Birds* (1963 and a Tomatometer reading of 100 percent). Somberly viewers are informed that when Hitchcock made the film people laughed at the premise: flocks of birds attacking humans en masse. A few years later, the narration notes, no one laughed when that actually occurred in a small town.

Perhaps more significant than the movie's antecedents are its theoretical progeny. No, not the various Tomato sequels, but genre-spoof movies such as *Airplane* (1990) and *The Naked Gun* (1988). More than a few movie buffs have suggested that *Attack of the Killer Tomatoes!*, for all its failings, was the spark that ignited this series of groaningly funny films.

As for the sequels, they may have a better reputation than the first film, although there are not enough reviews on rottentomatoes.com to merit a Tomatometer percentage for any of them.

IT CAN'T GET WORSE. CAN IT!

Return of the Killer Tomatoes! is noted for the presence of brand names. John Astin (of the television Addams Family) appears as the evil Professor Gangreen, a role he reprised in future films. George Clooney (who wasn't quite *The* George Clooney then) has a part whose primary significance seems to be a dialogue in which he plugs as many soft drink, beer and chocolate brands as possible. And bizarre product placement is a running gag throughout.

Killer Tomatoes Eat France! is noteworthy for a heavy metal rock and roll riot that takes place in the Louvre–no B movie should be without one. And the director's cut adds footage and jokes that were sorely needed the first time out.

Meanwhile, that other tomato movie, *Fried Green Tomatoes* (1990), fares better than its riper kin, earning a Tomatometer rating of 90 percent.

For Fried Green Tomatoes, rated 100 percent delicious, turn the page!

Fried Green Tomatoes ala Early Girl

Serves 4 to 6

5 green tomatoes

2 cups/473 mg flour

1 tablespoon salt

1 teaspoon freshly ground black pepper

2 eggs

1/4 cup/59 ml milk

3 cups/708 ml bread crumbs

1/4 cup/59 ml grated parmesan

2 tablespoons minced fresh parsely

zest of 1 orange

vegetable oil for frying

- Remove core, trim ends, and slice the tomatoes about 1/2 inch/1.25 cm thick. Spread out on rack and allow to stand for 15 to 20 minutes.
- In a wide bowl, combine flour, salt, and pepper. In a second bowl, whisk together eggs and milk. In a third bowl, combine bread crumbs, parmesan, parsley, and orange zest.
- Heat about 1/2 inch/1.25 cm oil in a wide, heavy skillet over medium high heat until hot but not smoking.
- Dredge tomato slices first in the flour mixture, then in the egg wash, then in the bread crumbs, coating both sides well. Fry in small batches (don't crowd), turning once, until golden brown on both sides. Return to rack to drain.
- Serve immediately.

Fried green tomatoes will make your mouth water, whether you're from the South or not.

Fried Green Tomato Casserole with Spring Garden Vegetables

Serves 6 to 8

- Preheat oven to 425°F/218°C.
- Cover the bottom of a greased casserole or baking dish with half of the fried green tomatoes, followed by a layer each of the spinach, peas, and black beans, using all of each.
- Cover with 2 cups of the tomato gravy. Top this with the feta cheese, then a layer of the remaining fried green tomatoes.
- Spread the rest of the tomato gravy evenly over the top and finish with a covering of the grated parmesan cheese.
- Bake at 425°F/218°C for 45 minutes to 1 hour, until top is golden and bubbly. Serve warm.

30 slices fried green tomatoes *(see recipe on page 140)*

1 pound/½ kg spinach, sautéed and drained

3 cups/708 ml cooked fresh peas

4 cups Tomato Gravy *(see recipe page 000)*

1 cup/236 ml grated parmesan cheese

1 cup/236 ml crumbled feta cheese

2 cups/473 ml cooked black beans

salt

pepper

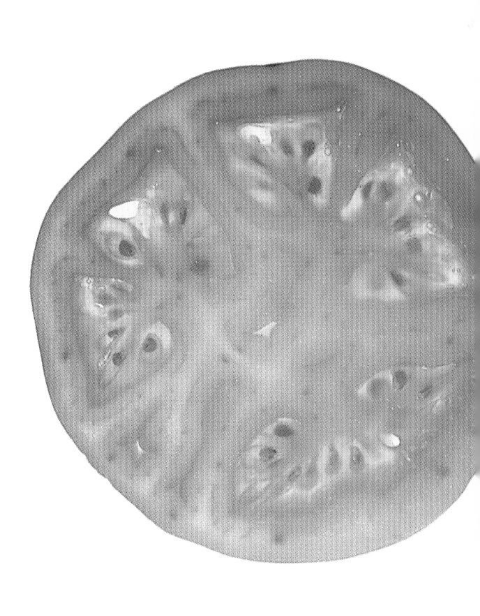

TOMATO KITSCH

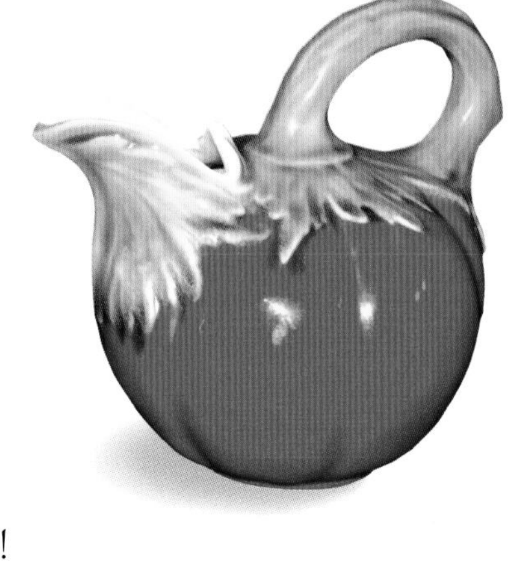

APPLES ARE TOO WHOLESOME,
TURNIPS SO HO-HUM,
BUT THE SAUCY-CHEEKED TOMATO
IS EVERYBODY'S CHUM

HE ANIMATES A PITCHER,
THEN SWITCHES ON "DE LIGHT."
GIVE HIM FEET AND WIND HIM UP
AND WATCH HIM DANCE ALL NIGHT!

THE TOMATO'S KITSCH POTENTIAL
IS NOTHING LESS THAN GREAT
BUT DON'T FORGET THAT WHERE HE'S BEST
IS SERVED UP ON YOUR PLATE!

No more bad verse, just terrific recipes when you turn the page

Spinach Potato Cakes with Tomato Gravy

Serves 6

Tomato Gravy

1/2 cup/118 ml finely diced onion
2 tablespoons olive oil
1 tablespoon plus 1 teaspoon white flour
2 cups/473 ml peeled, cored, and chopped tomatoes
water
1/2 teaspoon minced fresh basil
pinch dried thyme
1/2 tablespoon sugar
salt
pepper

- In a heavy saucepan, sauté the onions in the olive oil until translucent. Sprinkle flour over the onions and oil, and stir steadily until you create a light brown roux (see Note page 132). (A whisk is a good tool to use for this recipe.)
- Add tomatoes. You may need to add a little water as you cook to prevent sticking, but not too much. The mixture should resemble a very thick cream sauce. Stir in the basil, thyme, and sugar. Salt and pepper to taste. Simmer the mixture on low for 30 minutes, whisking frequently, until the tomato pieces have disintegrated.
- If you do not intend to use immediately, mixture may be kept covered in the refrigerator for 2 days. When you are ready to use, warm up in a microwave or in the top of double boiler.

Spinach and Potato Cakes

1 pound/1/2 kg spinach
1 tablespoon minced garlic
1 tablespoon olive oil
4 pounds/2 kg potatoes, cooked and grated
4 eggs, lightly beaten
2 cups/473 ml parmesan cheese
4 cups/946 ml bread crumbs
5 drops hot pepper sauce
zest of 3 lemons
1/2 teaspoon salt
fresh ground pepper

- Clean the spinach thoroughly and remove stems. Chop coarsely, then lightly sauté the spinach and garlic in the olive oil until spinach is wilted. Drain.
- In a mixing bowl, combine the spinach with the potatoes, eggs, cheese, crumbs, pepper sauce, lemon zest, salt, and pepper. When ingredients are thoroughly blended, form into 12 cakes approximately 3 inches/7.5 cm in diameter.
- Coat a heavy skillet with olive oil and, over medium high heat, fry the patties until golden brown and crisp on both sides. Be careful not to burn them.
- Serve two generously doused with tomato gravy for each person.

Spinach & Tomato Bread

Makes 2 loaves

- *1 cup/236 ml milk*
- *1 cup/23g ml packed, spinach leaves*
- *1/2 cup/118 ml parsley leaves*
- *1 tablespoon olive oil*
- *3 cups/708 ml bread flour*
- *2 tablespoons sugar*
- *1/4 ounce/7 gr yeast*
- *1 teaspoon salt*
- *3/4 cup/177 ml dried oil-packed tomatoes, drained*
- *1 cup/236 ml roasted sweet red peppers, drained*
- *2 tablespoons pine nuts*
- *1 garlic clove*
- *1 egg beaten with 2 tablespoons water*

- In a small saucepan on low, heat the milk until bubbles appear around the sides. Remove from heat.
- In a food processor, combine the milk with the spinach, parsley, and olive oil, using the chopping blade for 1 minute. Set aside while you prepare the flour mixture.
- In a large mixing bowl combine 2 1/2 cups/591 ml of the flour with the sugar, yeast, and salt.
- Test milk temperature and when it reaches 125°F/51.5°C, add it to the flour mixture, beating as you do. (If the milk mixture has cooled too much, very gently reheat it to the desired temperature.)
- Turn the dough out onto a flat, floured surface and knead the remaining flour into the dough. Place dough in lightly oiled bowl and cover with a damp towel. Allow to rise in a warm place until doubled in bulk.
- Preheat oven to 350°F/176°C. Lightly grease two cookie sheets.
- While dough is rising, prepare filling by combining the dried tomatoes, roasted peppers, pine nuts, and garlic in a food processor using the chopping blade, until coarsely pureed. Set aside.
- When the dough has risen, punch down and divide into two equal balls. On a floured surface, roll out dough into a rectangle and cover with half the tomato spread, leaving about 1 inch/2.5 cm uncovered on one edge. Starting with the covered edge, roll up the dough to make a long loaf. Place on baking sheet and make several slashes in the top for air vents. Repeat with the second portion of the dough and the remaining spread.
- Brush the tops of the loaves with the egg wash.
- Bake at 350°F/176°C for 30 to 35 minutes until tops are golden brown. Allow to cool before slicing.

A BROAD DEFINITION OF TOMATO

Hey, don't blame me. The term has been used since at least 1920 to describe a woman who was, well, ready for the picking. It was an especially handy handle in Hollywood from the 1930s to 1950s, although its meaning seems to have evolved over that time.

In the beginning, think Barbara Stanwyck: smart, sexy, shoulders square as any guy's but pretty as a pony and just as frisky. She's the recipient of one of the best lines in moviedom, delivered by costar George Brent in 1932's *The Purchase Price*. Embracing her, he says: "Ya daffy tomato, I'm bugs about ya!" Does that warm the cockles of your heart, or what?

Fourteen years later in *The Big Sleep*, Lauren Bacall, sings a different tune, this one with the lyric: "She's a real sad tomato/she's a foster valentine." (Not quite 50 years later, R.E.M. would reprise the concept in "Crush With Eyeliner," singing "I know you/I know you've seen her/She's a sad tomato/She's three miles of bad road/walking down the street.")

In 1955, Marilyn Monroe played "the tomato upstairs," a sexy single girl living upstairs from married man Tom Ewell in *The Seven Year Itch*. That's the movie that gave us the fabulous shot of Marilyn standing over a sidewalk grate in New York, her skirt billowing into the air–perhaps the best hot tomato pic of all time.

Two years later, though, Frank Sinatra gave tomato a new dimension when he took on the role of the philandering Joey Evans in *Pal Joey*. To get into his part, Sinatra created a lexicon of terms Joey would use to describe his chosen prey. "The way I figure it, broads can be divided into eight different classifications," Sinatra said.

Harvey Crabclaw,
M.M., c.1995 Photograph, candy tin, wood, metal stamp.

"The one to really watch out for is the tomato. She's a broad who's ripe for marriage," said Frank Sinatra

"There's the Mouse, the Tomato, the Beetle, the Quim, the Twist and a Twirl, the Gasser, the Barn Burner, and the Mish Mash.

"A Mouse is a cuddly broad. A Beetle is a flashy broad. One who makes with the sharp clothes. A Quim is a loose broad, one who's easy to pick up. A Twist and a Twirl is a broad who likes to dance. Of course, I suppose, everybody's heard of the word Gasser. Well, in Broadsville talk that means a dame who's a real looker, a knockout.

"Now take the Barn Burner: That's a broad with real polish and class. Who wouldn't dig her the most?

"As for the Mish Mash, she's a broad who's all mixed up. Of course, the one to really watch out for is the Tomato. She's a broad who's ripe for marriage."

A tomato sandwich? Barbara Stanwyck between Jack Benny and Jimmy Stewart.

The following recipes can have featured roles in your next dinner production.

Gumbo z'Herbes

Serves 12

9 tablespoons olive oil
2 cups/473 ml peeled and diced onion
2 cups/473 ml chopped bell peppers
1/4 cup/59 ml seeded and chopped hot peppers
1 cup/236 ml diced celery
2 cups/473 ml chopped pattypan squash
1 cup/236 ml corn kernels
1 cup/236 ml grated carrots
1 cup/236 ml sugar snap peas
1 teaspoon minced fresh basil
1 teaspoon fresh thyme leaves
2 teaspoons red pepper flakes
3 bay leaves
1 tablespoon minced garlic
1 pound/1/2 kg fresh greens, stemmed and chopped
5 tablespoons flour
3 cups/591 ml finely sliced okra
4 cups/946 ml peeled, chopped tomatoes
10 cups/2360 ml vegetable stock or water
1/4 teaspoon hot pepper sauce
salt
pepper
6 cups cooked rice

- In a soup pot, use 3 tablespoons of the olive oil to sauté onions, peppers, celery, squash, corn, and carrots for 5 minutes over medium heat.
- Add the peas, basil, thyme, red pepper flakes, bay leaves, garlic, and greens and sauté 5 more minutes. Set aside.
- In a heavy skillet, make a dark brown roux with 4 tablespoons of the olive oil and the flour. (See page 132 for information on making roux.)
- In a saucepan, use the remaining 2 tablespoons of olive oil to sauté the okra until it is golden brown. Stir in the tomatoes and set aside.
- Return the soup pot to low heat. Add the roux, stirring it into the vegetables to incorporate it throughout. Add the vegetables and stock, stirring to mix well. Add the okra and tomatoes, stirring well again. Add hot pepper sauce, salt, and pepper. Simmer for 2 hours.
- Serve hot over 1/2 cup cooked rice per person.

A Creole Lenten feast so full of garden goodness you won't notice there's no meat.

Tomato-Seed Bread

Makes 2 loaves

- *cornmeal for dusting*
- *3 1/2 cups/826 ml flour*
- *1/4 ounce/7 gr yeast*
- *1 cup/236 ml plus 1 tablespoon tomato juice*
- *1 tablespoon butter at room temperature*
- *1 teaspoon salt*
- *1 tablespoon sugar*
- *1 egg, slightly beaten*
- *caraway or poppy seeds*

- Preheat oven to 375°F/190.5°C. Prepare two baking sheets by dusting them with cornmeal.
- In a large mixing bowl, combine 1 cup/236 ml of the flour with the yeast.
- In a saucepan, warm the tomato juice, butter, salt, and sugar to 125°F/51.5°c. Pour into the flour and beat for 30 seconds.
- Add additional flour 1/2 cup/118 ml at a time, incorporating well after each addition. You may not need all the flour. You want to get a dough that holds together well and is not soggy but is moist and pliable.
- Turn the dough out on a floured surface and knead for 10 minutes.
- Place in oiled bowl and put in warm place, cover with a damp cloth and let rise for 1 hour.
- Divide dough in half and shape into long slender loaves. Place on baking sheets in warm place and let rise until doubled, about 45 minutes.
- Brush the top of each loaf with the egg wash and sprinkle the seeds over it. Press them in very, very lightly with the palm of your hand. Slash diagonal air vents in the top of each loaf.
- Bake at 375°F/190.5°C for 30 minutes, until crust is golden. Allow to cool before slicing.

This savory bread is the perfect foil for the many soups and stews you can make with tomatoes.

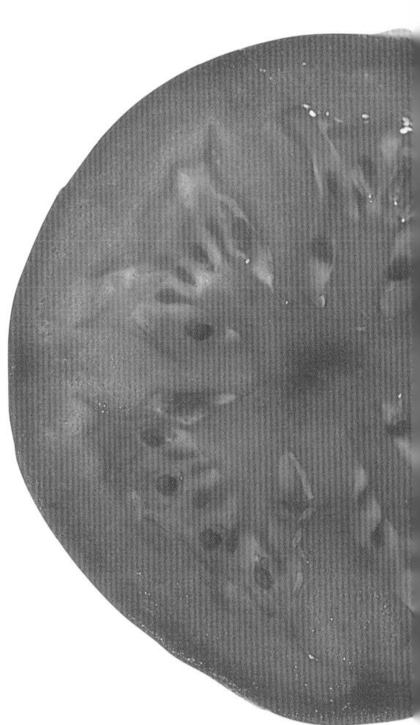

Minestrone

Serves 12

- *1 cup/236 ml diced onion*
- *1 cup/236 ml diced carrot*
- *2 cups/483 ml diced celery*
- *1/4 cup/59 ml olive oil*
- *4 cups/946 ml peeled and chopped tomatoes*
- *1 cup/236 ml green beans in 1-inch/2.5-cm lengths*
- *1 cup/236 ml sliced zucchini*
- *3 cups/591 ml chopped spinach*
- *2 cups/473 ml peeled and diced potatoes*
- *3 cups dry elbow macaroni*
- *2 cups cooked kidney beans*
- *5 cups/1180 ml vegetable stock or water*
- *1 teaspoon salt*
- *2 teaspoons chopped parsley*
- *2 teaspoons miso paste*

❖ In a soup pot on medium heat, cook the onions, carrots, and celery in olive oil until tender. Add the tomatoes, green beans, and zucchini and sauté for 5 minutes.

❖ Add the spinach, potatoes, dry macaroni, kidney beans, and vegetable stock. Simmer for 30 to 40 minutes, then add salt, parsley, and miso paste. Simmer for another 10 minutes, then serve.

This distinctive regional dish of Italy
gets its body and soul from tomatoes.

FROM TOMATINA TO TOMATO ART:

FESTIVALS AROUND THE WORLD

Spain has its gazpacho; France, its bouillabaisse; Kentucky, burgoo; Livorno is known for its cacciucco. Just as every region seems to have a classic variation on the tomato soup theme, so does the world sport a variety of festivals that celebrate the tomato, each with a distinctly different flavor.

WOULD YOU LIKE SOME GRATED MANCHEGO ON THAT?

Certainly the most outrageous is Spain's La Tomatina. Every year on the last Wednesday in August, the small town of Bunol, only 30 miles from Valencia, plays host to the world's largest food fight. There beneath an imposing medieval bell tower, in streets so narrow they might pass elsewhere for alleyways, more than 20,000 revelers from around the world gather to be pelted by overripe tomatoes.

That's right, pelted by tomatoes—approximately 150,000 hauled there in huge dump trucks. From 11 a.m. to 1 p.m. folks are allowed—nay,

encouraged—to grab a tomato and fling it on a friend or perfect stranger. It's not an utter free-for-all. There are rules. Tomatoes are to be squished first before hurling, presumably so you won't prove Mama right and put out someone's eye. And nothing, absolutely nothing, but tomatoes can be hurled. You are also forbidden from ripping off another's clothing, although from pictures it appears there is no rule against discarding your own shirt. When the two hour tomato splat ends, the sauced are invited to shower down by the river in makeshift stalls that have been set up for this purpose.

The festival began informally in 1944 when several friends hanging out in the town square started a food fight. Some stories say there was a reason, some say not; but for the most part, everyone seemed to find it a heck of a way to spend a slow Wednesday morning, and an annual event was born.

As the party began to catch on and adventurous tourists arrived, the town decided to expand the festival into a weeklong event that also celebrates Bunol's patron saint. Now there are parades and fireworks, and the night before the great tomato fight, the streets are filled with simmering pans of paella, and the wine flows.

ONLY LA PARMIGIANA, PLEASE

Of course, they would not dream of throwing the tomatoes of Sardinia, so delicious are they reputed to be. Instead the town of Zeddiani on this Italian island has a festival every August in which the tomato is prepared according to recipes that have been passed down from generation to generation. The tomatoes are then served in a communal meal where folks gather not only to eat but to share recipes and culinary ideas. The event also includes a gastronomical food fair that highlights products of the region.

LE PETIT PRINCE DE TOMATE?

Meanwhile, at the Chateau de La Bourdaisiere, about 50 miles outside of Paris in the Loire valley, you are invited mid-September to experience everything to do with the tomato, except ketchup, of course. Two brothers who are princes, one of them a gardener, own the chateau. They have turned the castle and its grounds into a bed and breakfast with a vast tomato garden of some 500 unusual varieties.

("With 10,000 varieties of tomato, why should they all be round and red with no taste?" asks Prince Louis-Albert de Broglie.) There you will find a shop which features gardening implements and clothing and dozens of products made from estate tomatoes, including tomato jam, tomato face cream, and tomato soap. The annual tomato festival features tastings of the estate crop and cooking demonstrations from regional chefs.

A PLURALITY OF FESTIVITIES

In the United States more tomato festivals seem to crop up each year. Many of them revive festivals from the bygone days of local canning and commercial tomato production. Often these festivals began in the 1930s or 1940s during the boom and ended in the 1960s when companies began to consolidate and plants moved to California. In recent years, however, many towns have decided to capitalize on their history and to celebrate local homegrown tomatoes with contests for largest, ugliest, tastiest, and the like. These local festivals can be great places for sampling many varieties of tomato fresh from the vine.

Reynoldsburg, Ohio, calls itself the birthplace of the commercial tomato and the festival there each September honors not only the product but a local seed and plant merchant who did much to bring it into being. Alexander Livingston developed or improved 13 major varieties of tomato from 1870 to 1893, including the Paragon, the first variety developed for commercial production. Livingston died in 1898 but is commemorated in a Hall of Fame in Reynoldsburg, and the literal fruits of his labors have been celebrated there since 1965. There are some 40 contests for tomatoes and free tomato juice (the official state drink of Ohio) is served, starting on Wednesday after Labor Day.

Though the Kendall-Jackson Wine Estates' Heirloom Tomato Festival in Fulton, California, is a relative newcomer to the scene, having begun in 1997, it celebrates the beauty of old tomato varieties. Some 175 different tomatoes grown in the organic gardens of the estate are available for sampling at the event, which takes place in the winery's extensive gardens. Samples of tomato dishes from 50 of the region's justly celebrated restaurants and food purveyors are also on tap, and there is a wine tasting as well. The usual contests for biggest and best, and more, are a part of the event, with a focus on heirlooms. But most unusual is the Tomato Art Exhibit. A juried art show featuring the tomato as subject is hung annually at the Wine Center in Santa Rosa, culled from entries from around the world.

A couple of artful
recipes followed by the
Tomato Art Gallery

Red Snapper with Green Tomato & Blackberry Sauce

Serves 4

1 green tomato

1 tablespoon fresh lemon juice

1 teaspoon minced lemon zest

3/4 cup/177 ml sugar

pinch of ground cinnamon

pinch of ground nutmeg

1/4 cup/118 ml water

1 pint fresh blackberries

salt

4 red snapper filets 5-ounce/150-gram each

olive oil

- Preheat oven to 450°F/232°C.
- Core the green tomato and puree it in a blender or food processor. In a non-stick saucepan, bring the tomato puree to a low boil on medium heat.
- Add the lemon juice and zest, sugar, cinnamon, nutmeg, and water. Lower the heat and let simmer until mixture is the thickness of a rich marinara.
- Remove from heat and gently stir in blackberries. Add salt to taste. Set aside while you prepare the fish.
- Lightly oil and salt both sides of the snapper filets. Place on rack over pan and bake in 450°F/232°C oven until fish is flaky, about 10 minutes per inch/2.5 cm of thickness.
- Place filets on plates and top with sauce.

There's nothing fishy about this bold combination of fine flavors.

Green Tomato End of Harvest Soup

Serves 12

4 carrots, diced

4 celery stalks, sliced

2 large onions, diced

2 tablespoons olive oil

2 zucchini, sliced

2 butternut squash, peeled, seeded, and chopped

2 cups/473 ml cooked beans (any available)

3 ears of corn, shucked and cut from the cob

4 green tomatoes, cored and diced

3 cups/591 ml chopped greens (kale, beets, turnip, collard)

3 potatoes, peeled and cubed

1 teaspoon fresh thyme leaves

3 tablespoons minced scallion

6 cups/1415 ml vegetable stock or water

Salt

Pepper

- In soup pot over medium heat, sauté the carrots, celery, and onions in olive oil until they begin to turn tender.
- Add the zucchini, butternut squash, and beans, and continue cooking until the mixture starts to brown.
- Add the corn, green tomatoes, greens, potatoes, thyme, and scallion and cook for 2 minutes.
- Add the vegetable stock (or water) and cook until the potatoes are soft.
- Salt and pepper to taste. If more water is needed to achieve the consistency of thick soup, add it and continue to cook until heated through. Serve hot.

TOMATO GALLERY

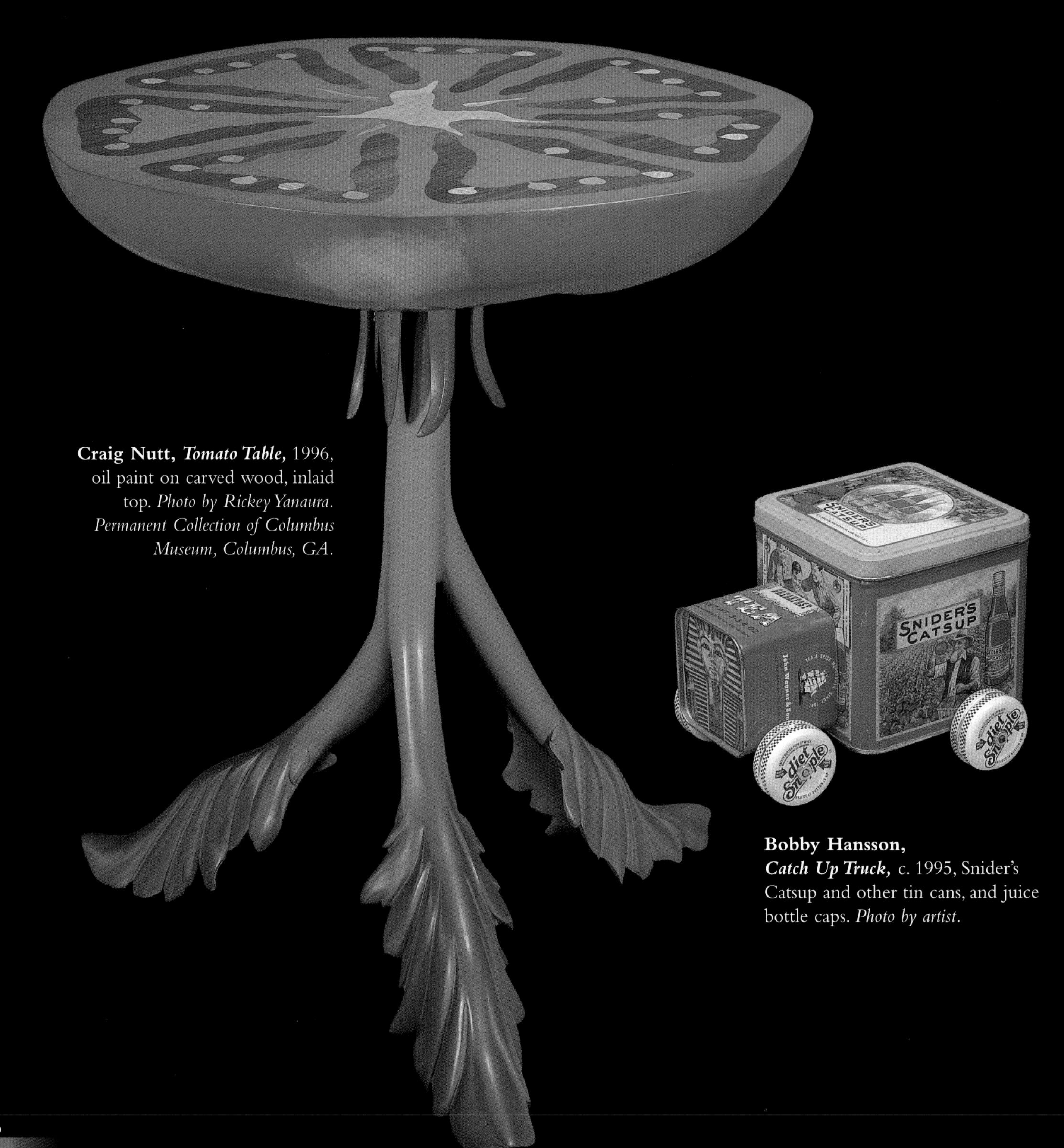

Craig Nutt, *Tomato Table,* 1996, oil paint on carved wood, inlaid top. *Photo by Rickey Yanaura. Permanent Collection of Columbus Museum, Columbus, GA.*

Bobby Hansson,
Catch Up Truck, c. 1995, Snider's Catsup and other tin cans, and juice bottle caps. *Photo by artist.*

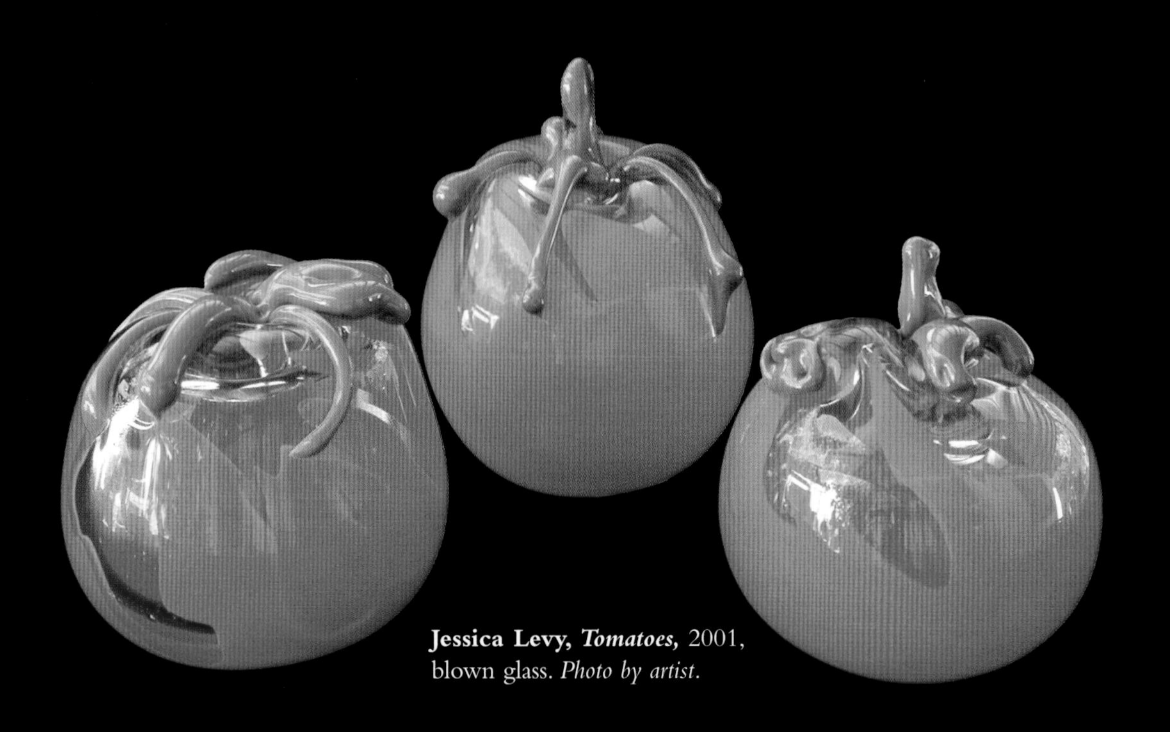

Jessica Levy, ***Tomatoes,*** 2001,
blown glass. *Photo by artist*.

Barnee Alexander, ***Who's Minding the Garden***

Julie and Tyrone Larson, ***Pasta B[illegible]***, 2003

TOMATO GALLERY

Susan Stryker,
Still Life, 1963,
watercolor.
Photo by Dana Irwin

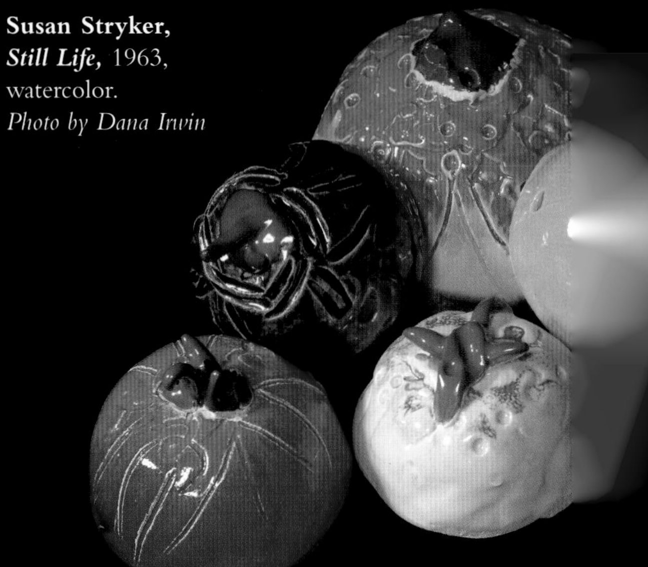

Marjorie Fitterer,
Cluster of Tomatoes, 2002, ceramic.
Photo by Rossen Townsend

Anonymous, *Get Your 'Mater Running,* tomato cans.
Photo by Stan Rosen.

Tina DiCicco, *Cluster,* 2003, oil pastel.
Photo by artist.

Mariel Green, *Variations on*
Photo by artist.

TOMATO GALLERY

Bobby Hansson, ***Kitchen Items,*** various tomato and other tin cans. c. 1995. *Photo by artist.*

Barnee Alexander, ***California Bearchus Enjoys Tomatoes,***

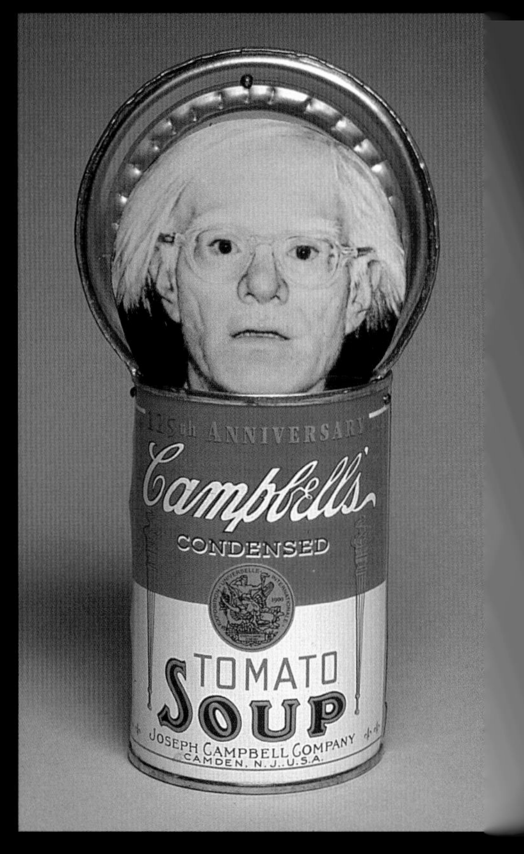

Harvey Crabclaw, ***Saint Andy Rising,*** tin ca

Susan Cornelis, *It's in the Bag,* 2002, watercolor. *Photo by Bob Cornelis.*

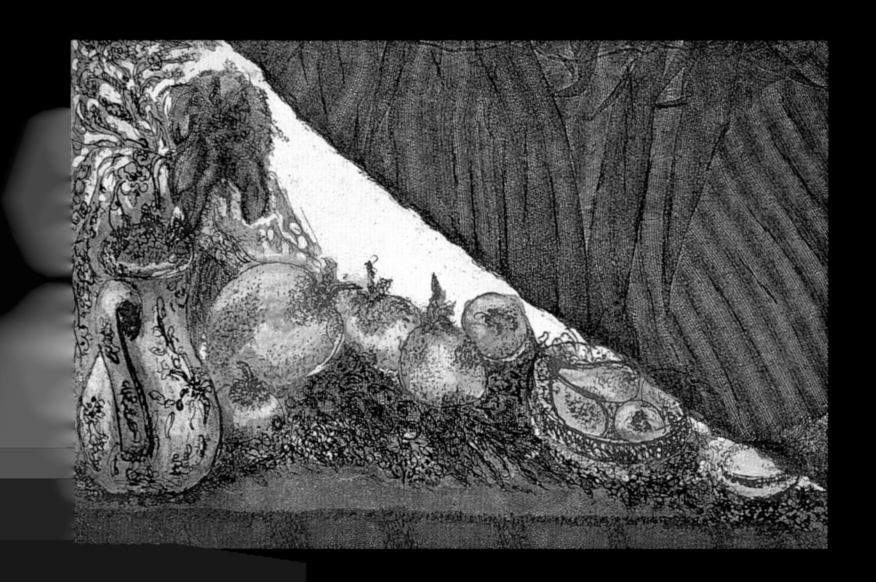

TOMATO GALLERY

Mary Ann Henderson, *Heirlooms,* 2002, pastel. *Photo by D. Gregory Henderson.*

Jack A. Lutzow, *Vibrancy,* 2000, watercolor. *Photo by artist.*

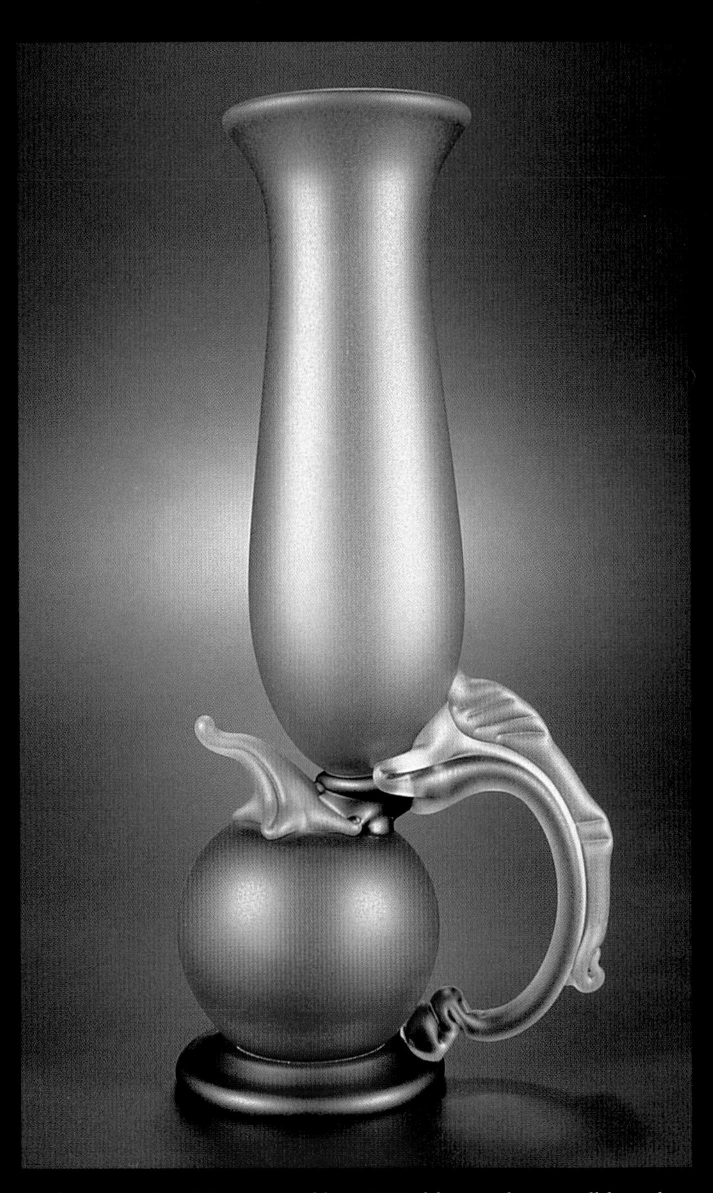

Robert Levin, *Tomato Goblet,* 2002, blown glass, sandblasted and ACI. *Photo by artist.*

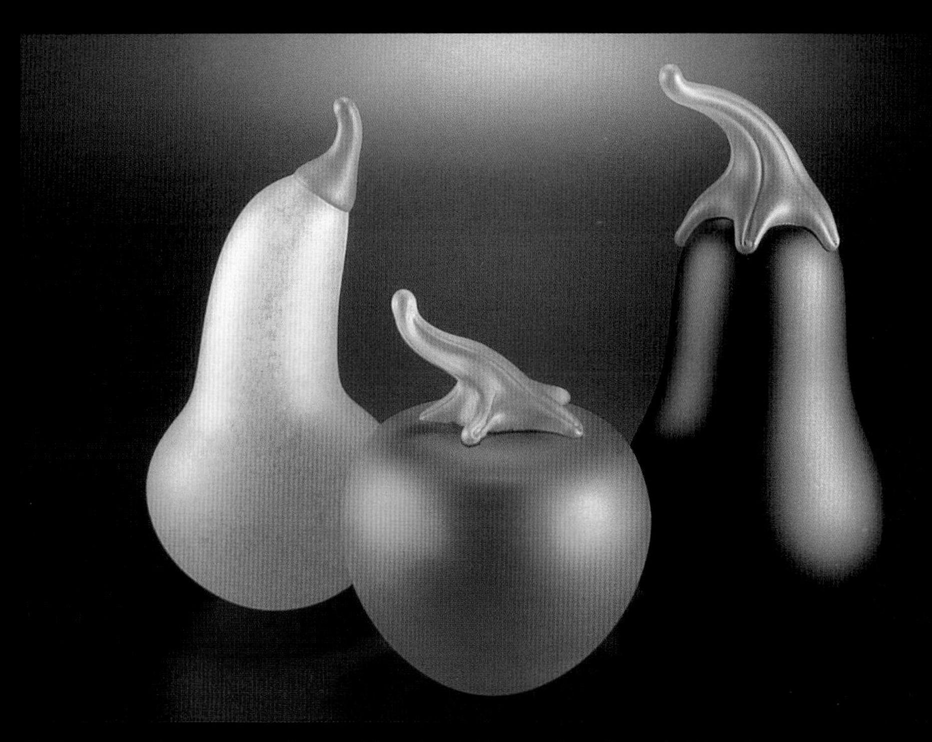

Robert Levin, *Fruit and Vegetables,* 2002, blown glass, frosted. *Photo by artist.*

Gus Riley, *Plum Tomatoes and Green Cloth,* 1997, oil on board.
Photo by Beach Photo

Acknowledgments

Telling the true, unexpurgated story of the tomato has been a genuine group effort. As already noted, John and Julie Stehling of the Early Girl Café came up with delicious, cogently written recipes for In Praise of Tomatoes. They also provided comfort and food for the crew that put this book together on more than one occasion. And Barbara Ciletti, took time from her multiple projects to supply a wealth of helpful garden tips. Thanks for making us welcome both at the table and in the garden.

Mere words cannot describe the allure this fragile fruit has had on us for centuries, however. To that end, art director Dana Irwin filled this book with gorgeous images of the tomato in all of its many guises, from palate pleaser to Hollywood starlet to the saucy waitress who points the way to recipes. Dana is such a tomato fan that she not only grew some for the photos, but also contributed the lovely watercolors in the garden section and wrote the delightful limerick on page 86.

Making Dana's vision come to life was Sandra Stambaugh, Asheville, North Carolina photographer. Sandy's patience and craftsmanship are evident in the gorgeous images of tomatoes on the table, tomatoes on the vine, tomatoes in the pantry, and tomatoes at the farm. Thanks to her artistry, this book is a splendid visual feast.

We have editorial assistant Nathalie Mornu to thank for the riot of classic, kitsch, and fine art tomato images that chronicle the tomato's lively life as a cultural icon in this book. Nathalie not only brought great tomato pictures to the table, but unflagging enthusiasm. And Rebecca Lim stayed steady and focused when the project seemed to be growing as fast and unpredictably as an indeterminate tomato vine.

Assistant Art Director Lance Wille brought a keen eye, deft hand and limitless good humor to the design phase of this book. A tip of the snap-brim porkpie to him.

A host of tomato-philes stepped up to the plate when we went in search of images of this fine fruit in art, in kitsch, in the garden and through history.

The images of tomatoes on pages 78-85 in The Compleat Chart for Cultivating & Cooking Tomatoes, appear courtesy of W. Atlee Burpee & Co.; and Chip Hope of Appalachian Seeds in Flat Rock, North Carolina, who also contributed the images of heirlooms on pages 54 & 55.

We would also like to thank these contributors:

Stuart Alexander (Australia) PTY Limited for permission to use the postcard image on page 24 from the private collection of Eddyvale Enterprises, Australia;
Antiques in Paradise for the images of Royal Bayreuth tomato pottery on page 145, photos by Sherry Ann DeGrave; Donna Diken of Instant Living for the tomato light switch, page 145; Sam Earle for tomato can labels on pages 17 and 127-129, with the exception of the Lark Brand label (see below); H. J. Heinz Company, L.P. (Heinz), for permission to use the advertisements on pages 119, 136 and for the black and white photos of ketchup production on page 125; Corinne Kurzmann of Diggin Art for the dish cloth on page 145; The Library of Congress for the chart on page 14, and images of historical advertisements and posters on pages 65 and 126; Christine McClaine of Rose Internet Auctions for the pitcher on page 144; Franchi Sementi, S.p.A of Italy for the seed packet on page 63; Kendall Jackson Wine Estates, Ltd. for photos of the estate's heirloom tomatoes and festival on pages 158-161; Seneca Foods Corporation for the Stokely-Van Camp advertisement, page 136; TinManTinToys for the wind-up tomato on page 144; Tomatina Tours for photos of La Tomatina by Jason Goodyear and Dan Lefebvre on pages 158 & 159; The Webb family, with permission of Jason Gimble for the Lark Brand tomato can label on page 3.

Tomato Gallery Artists

Barnee Alexander grows, roasts, freezes and immortalizes tomatoes in art in Santa Rosa, California.

Painter Tina DiCicco lives in Rohnert Park, California.

Watercolorist Susan Cornelis lives in Sonoma County and is also the illustrator of a children's book, *The Great Hiss*.

Harvey Crabclaw is an itinerant folk artist whose studio is a school bus.

Marjorie Fitterer first cast "Cluster of Tomatoes" on page 168 over round glasses before hand-forming. She lives in San Francisco, California.

Loretta Frediani is a well-known Sonoma County, California folk artist who often works in wood.

Mariel Green lives in Santa Rosa, California.

Bobby Hansson has been making art from found objects since 1955. He is the author of *The Fine Art of the Tin Can*, (Lark Books) and lives near Rising Sun, Maryland.

Mary Ann Henderson's "The Tomato Queen" on page 168 is a self portrait. She lives in Saratoga, California.

Julie and Tyrone Lawson, a pottery team since 1966, work and have a shop in Asheville, North Carolina.

Robert Levin is an internationally known glass artist who lives and works near Burnsville, North Carolina. His work can be seen at robertlevin.com.

Jessica Levy is a glass artist living in in Cotati, California.

Jack A. Lutzow lives in Glen Allen, California.

Acclaimed woodworker Craig Nutt creates furniture and sculpture in his studio near Nashville, Tennessee. You can see more of his work at craignutt.com.

Florida artist Gus Riley's work is featured in the Lark Book *Design!*.

Rakshika Thakor lives in Santa Rosa, California.

Index